D0383162

EUGENIE CLARK

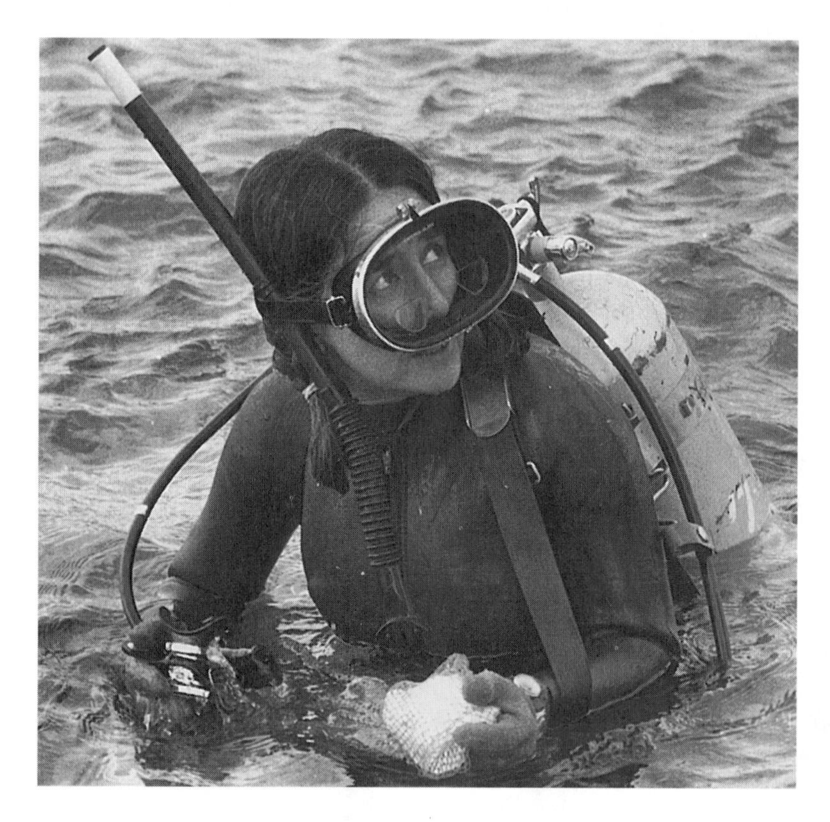

Eugenie Clark, 1975

Butts, Ellen.
Eugenie Clark :
adventures of a shark
2000.
33305015078508
LA 03/30/01

EUGENIE CLARK

ADVENTURES OF A SHARK SCIENTIST

by Ellen R. Butts
and Joyce R. Schwartz

LINNET BOOKS ··· 2000

© 2000 Ellen R. Butts and Joyce R. Schwartz.
All rights reserved.
First published 2000
as a Linnet Book, an imprint of The Shoe String Press, Inc.,
2 Linsley Street, North Haven, Connecticut 06473.

Library of Congress Cataloging-in-Publication Data

Butts, Ellen.
Eugenie Clark : adventures of a shark scientist / by Ellen R. Butts and Joyce R. Schwartz.
p. cm.
Includes bibliographical references (p.).
Summary: Describes the life and career of ichthyologist Eugenie Clark, who first became interested in fish at the New York Aquarium and went on to become an expert on sharks.
ISBN 0-208-02440-9 (lib. bdg. : alk. paper)
1. Clark, Eugenie—Juvenile literature.
2. Ichthyologists—United States—Biography—Juvenile literature.
3. Sharks—Research—Juvenile literature. [1. Clark, Eugenie.
2. Ichthyologists. 3. Women—Biography. 4. Sharks.
5. Zoologists.] I. Schwartz, Joyce R. II. Title

QL31.C56E84 1999
597'.092—dc21
[B] 99-044957

The paper in this publication meets the minimum requirements of American National Standard for Information Sciences—Permanence of Paper for Printed Library Materials, ANSI-Z39.48-1984.∞

Designed by Sanna Stanley

Printed in the United States of America

To my father, Seymour Rubinstein, scientist and role model.
E.R.B.

To my husband, Richard, for his love and support.
J.R.S.

CONTENTS

AUTHORS' NOTE

Despite her hectic schedule, Dr. Eugenie Clark provided invaluable assistance by granting us an interview and answering our many follow-up questions. She furnished details that confirmed our research, and told us colorful stories that added an extra dimension to our writing.

Best of all, it's fun to spend time with Genie. Although she has achieved international renown as an ichthyologist and led a life filled with remarkable adventures, she retains her warmth, sense of humor, unpretentious attitude and unending curiosity about what each day will bring.

Unless otherwise noted, all pictures in this book are from Genie's personal collection and are used with her permission.

CHRONOLOGY OF EUGENIE CLARK'S LIFE

400 MILLION YEARS AGO
—sharks first appear on Earth

1922—Eugenie Clark, the future "shark lady," is born on May 4

1931—first visits New York Aquarium

1934—William Beebe and Otis Barton descend over 3,000 feet into the ocean in a bathysphere

1938—Genie enters Hunter College

1941—Japan attacks Pearl Harbor; U.S. enters World War II

1942—Genie graduates from Hunter College; marries Roy Umaki

1943—Jacques-Yves Cousteau and Emile Gagnan design the compressed-air Aqua Lung, making scuba diving available to everyone

1946—Genie earns master's degree in zoology from NYU; works with Carl Hubbs at Scripps Institution in California; learns to dive

1949—studies fish in the South Seas for the U.S. Navy

1950—marries Ilias Konstantinou

1951—studies fish in the Red Sea on a Fulbright scholarship; earns Ph.D in zoology from NYU

1952-1958—Hera, Aya, Tak and Niki are born

1953—*Lady with a Spear* is published

1955—Genie becomes founder and executive director of Cape Haze Marine Laboratory

1958—conducts groundbreaking experiments in shark training

1959—sets a women's world record for freshwater diving

1967—leaves Cape Haze; is divorced from Ilias

1968—becomes a professor at the University of Maryland

1969—*The Lady and the Sharks* is published

1972—Genie studies the Moses sole in the Red Sea; begins writing for *National Geographic* magazine

1973—first observes "sleeping" sharks

1979—begins conservation efforts in Red Sea

1981—takes first ride on a whale shark

1987—becomes chief scientist of the Beebe Project and makes the first of many submersible dives

1991—*The Desert Beneath the Sea,* a children's book about sandfish, is published

1992—Genie retires officially from the University of Maryland but continues teaching, leading research expeditions and writing scientific and popular articles as a Professor Emerita and Senior Research Scientist

1999—teaches last class at University of Maryland but continues to dive and study sandfish and sharks

2000+—will lead research expeditions to the South Pacific and La Paz, Mexico; will maintain labs at the University of Maryland and the Mote Marine Laboratory

1
AN UNUSUAL CHILDHOOD

The diver stared spellbound at the enormous monster swimming rapidly toward her. Its huge mouth was opened wide enough to swallow her in one gulp. For the first time Dr. Eugenie Clark was face-to-face with a live whale shark. She could have moved out of its path; she could have let it pass her by. But here was an adventure she couldn't resist.

As the shark veered suddenly to the right, Dr. Clark ducked around the enormous *pectoral fin* on its chest, then pulled herself upward and grabbed the *dorsal fin* on its back. Again and again the giant plowed up through the water to the ocean's surface, then plunged to the depths below. Dr. Clark's arms and hands became tired and sore, and soon she had to loosen her grip. But determined not to give up, she pulled herself onto its back, riding the shark like a cowgirl on a bucking bronco. Her steed, however, was the size of twenty-five horses and its rough, sandpapery skin rubbed against her bare legs, causing them to bleed. Dr. Clark shifted again and lay on her stomach, then slid slowly all the way down to grasp the shark's tail. Now, instead of being jerked up and down, she was whipped from side to side. Her face mask, scuba tank and flippers soon came loose. Her life was in danger; she realized that she'd have to let go. Though she had traveled only a short distance on her fantastic ride, she had traveled a long way in a life filled with adventure.

Eugenie Clark, known as Genie, came from a family of strong-willed women. Her grandmother, Yuriko, had had a traditional Japanese upbringing. As a young woman, she was married off to a much older man, a doctor in the Imperial Court of Japan. After several years, the doctor died, leaving Yuriko with two small children, a daughter named Yumiko and a son, Boya. She remarried a good-looking musician who soon accepted a job offer in Seattle, Washington. Yuriko, Yumiko and Boya moved to the United States with him around 1910. When that marriage failed, Yuriko and her teenaged children moved across the country to New York City.

In 1919, Yumiko, who was an excellent swimmer, applied for a job as an instructor at a private pool. She was hired by its manager, Charles Clark. He fell in love with her and soon they were married. On May 4, 1922, Genie was born. Two years later her father died, leaving Genie's mother and grandmother to raise her on their own. The three of them lived in a tiny apartment in Queens, a borough of New York City. With no husband to provide for the family, Yumiko had to work long hours to earn a living.

Yumiko, Grandma Yuriko and Uncle Boya all loved to swim and, often on summer weekends, the whole family went to the beach on Long Island. Before Genie was two, Yumiko had taught her to swim. They spent many happy hours playing in the waves together. When she grew older,

Eugenie Clark's grandmother Yuriko, left, with her children Boya (front) and Yumiko (Genie's mother) back, with their nurse in Japan, 1900.

Genie loved to watch as her beautiful mother swam gracefully to shore. As she came out of the water, Yumiko would remove her bathing cap, letting her long black hair tumble down her back. She looked like an exotic Japanese pearl diver. Genie was proud when people realized that Yumiko was her mother.

At Woodside Elementary School, Genie was the only student from a Japanese family. At that time, there were very few Japanese people living in the eastern United States, so their customs and foods seemed strange to her classmates.

Yumiko Clark, in traditional Japanese kimono and hair style, holding baby Genie.

Meals at Genie's house included a lot of fish because her mother and grandmother cooked traditional Japanese food. Genie knew more about fish—especially eating them—than any of her friends. She enjoyed shocking the other children by describing how she ate rice and seaweed for breakfast and raw fish, called *sashimi*, and ground shark cakes for dinner. Even her school lunches weren't like anyone else's—she brought *sushi*, a roll of raw fish and vegetables wrapped in sticky rice, made fresh every day by Grandma Yuriko.

Genie liked *sushi* and didn't mind being teased about her strange lunches. But when the kids picked on her for being Japanese, she became angry and got into fights. Her grades in conduct were often bad; otherwise, she was a straight "A" student. One time, Genie won an art contest. She was very proud when her picture was hung in the school's auditorium. But someone wrote "the Jap" across it, and when Genie found out who was guilty, she tried to beat him up. Both of them were sent to the principal's office and, as usual, she got an "F" in conduct.

Fourth grade was an unforgettable year. Genie made a new friend named Norma. She and Norma promised to be best friends forever—and they still are. And Genie loved her teacher, Miss Riley, because Miss Riley cared about her and encouraged her interest in animals. Miss Riley wanted everyone in her class of city-raised children to see nature's wonders for themselves so she took them on several outdoor field trips. When she arranged a special field trip for a few select students, Genie was anxious to be included. Miss Riley said she could go—if she got an "A" in conduct. It was very difficult for Genie to control her temper, but Norma helped by constantly reminding her of the trip. Genie earned the "A" and her reward.

Yumiko worked at a newsstand in the Downtown Athletic Club, a building set among the towering skyscrapers and dark narrow streets at the southern tip of Manhattan, the oldest borough of New York City. She had to work on Saturday mornings. When Genie was old enough, she rode the subway to work with her mother and waited while the morning hours slowly passed. Then after work, Yumiko would usually take Genie for lunch at their favorite Japanese restaurant, owned by a good friend named Nobusan.

While Yumiko worked, it would have been fun for Genie to spend time at the water's edge, gazing out at the boats in

New York harbor and the majestic Statue of Liberty rising from Liberty Island. But she was only nine. Her mother didn't want her on the streets of Manhattan alone. Fortunately, the New York Aquarium was nearby. Yumiko decided that it would be safe for Genie to spend the mornings there instead. Neither of them imagined that this convenient arrangement would lead to a lifelong passion and a remarkable career.

The Aquarium's glass tanks were filled with strange and colorful fish, including a big shark. Genie was fascinated. Every Saturday she stared at the fish and pretended that she was walking among them on the ocean floor. She wondered how they could swim without arms and legs. After many Saturdays of watching, Genie realized that they used their powerful body muscles to swim and their fins and tails to steer. Later in her life, Genie remembered, "I never tired of watching the fish—from the streamlined, fast-swimming ones gliding back and forth in their long tanks like caged tigers, with barely a perceptible movement to explain their swift motion, to the sluggish, bottom-creeping forms that seemed to exert enormous effort through their whole bodies to inch along."

Nine-year-old Genie began to read books about sharks and other fish and learned all she could about the Aquarium's collection. She wanted to share the wonders of the fish world with anyone who'd listen. When the weather was cold and rainy, a group of homeless men—poverty-stricken by the collapse of the American economy that began the Depression in 1929—would gather in the Aquarium to keep warm. On Saturdays, they got something extra. Genie would lead them on tours, explaining what she had learned about fish and answering their questions. The guard thought

ANIMALS OF THE OCEAN

The ocean is home to all kinds of creatures. Some are *vertebrates*, animals with backbones. The largest group of vertebrates in the ocean are the fish, but there are also mammals, like whales and seals, and reptiles, like sea turtles and sea snakes. Each group has its own special characteristics. Fish have gills and take oxygen from the water, while reptiles and mammals have lungs and breathe oxygen from the air. Both fish and reptiles are cold-blooded; most are covered with scales and lay eggs. Mammals are warm-blooded, have fur or hair and give birth to live babies and nurse them.

There are many more *invertebrates* than vertebrates in the ocean. Invertebrates have soft bodies that are usually protected by a hard outer covering called an exoskeleton. They range in size from the 60-foot-long giant squid to the microscopic plankton, tiny creatures that float near the water's surface. Shellfish like clams, lobsters and snails are invertebrates; so are anemones, sea stars and corals. Corals build reefs that are home to many other animals. Coral reefs also provide a colorful and fascinating habitat for divers to explore.

the men were "bums," but to Genie they were students.

Genie longed for an aquarium of her own. Although Yumiko didn't share her daughter's fascination with fish, she was happy that Genie had such an interesting hobby. One wintry Saturday afternoon, Yumiko took her Christmas shopping after lunch. Both were filled with excitement as they walked to a nearby pet store. Genie picked out what she wanted—a 15-gallon fish tank. Then she and her mother added gravel and stones and some aquatic plants to fill its bottom. Finally it was time to choose the fish. At last they decided on pale green swordtails and black-speckled red platies, veil-tailed guppies and a pair of angelfish. It was hard to stop and before they'd finished, Yumiko had spent enough money for many Christmases and birthdays to come.

Now that she could observe the fish in her own home, Yumiko began to share her daughter's enthusiasm. Often she spent her lunch break at the pet shop, using her lunch money to buy a fishy addition for the aquarium. Genie also visited the pet store often, and spent all of her weekly allowance money there. Soon hundreds of fish were crowded into her big aquarium and a growing number of smaller tanks. Genie became the youngest member of the Queens County Aquarium Society, learning how to keep exact records—the date she got each fish, its scientific name and what happened to it.

As the Great Depression wore on, the Clarks had a hard time paying for basic necessities. Genie no longer received a weekly allowance, but she somehow managed to add to her animal collection. Toads, salamanders, snakes and even a small alligator shared the small Clark apartment along with Genie's fish. One of her favorite pets was a large black snake she named Rufus. Rufus often got Genie into trouble by escaping from his cage in her room and turning up in unexpected places. Some mornings, Genie was late for school because her grandmother wouldn't let her leave until she had captured Rufus and put him back in his cage.

Genie remained passionate about fish all through high school. In her English classes, no matter what topic the teachers assigned, Genie usually managed to write something about fish. Biology was her favorite subject because facts about animal *anatomy* and *physiology* helped her take better care of her pets and understand their behavior.

Genie discovered the work of William Beebe, a zoologist, world explorer and writer who built a marine laboratory on an island in the Bermudas. He became famous for plunging over 3,000 feet down into the ocean in a hollow iron ball, called a *bathysphere*. Gazing out from the safety of the bathysphere, Beebe was able to study sea creatures that had never been seen before.

"I told my family I would like to go down into the sea and be like William Beebe," Genie remembers. "They said maybe you can take up typing and get to be a secretary to William Beebe or somebody like him. I said, 'I don't want to be anybody's secretary!'" Her mother and grandmother were willing to encourage fish as a hobby, but they just couldn't imagine a woman making a career of studying fish and diving deep into the ocean. In the society of the 1930s, educated women could expect to teach or work in an office, then get married and have babies. At least, that's what most women did.

But Genie was different: She was determined to turn her hobby into a career in *ichthyology*, the study of fish. She would have to earn a college degree and then a doctorate in *zoology*, the study of animals. But a private college education was expensive, and Genie's family couldn't afford the fees. Fortunately, New York City had several excellent colleges that were free to residents. Genie entered Hunter College in 1938.

Genie took every zoology course Hunter offered. Many required long hours in the laboratory. She especially enjoyed the anatomy labs where the students *dissected* small animals, such as cats and guinea pigs, and studied the way their bones, muscles and internal organs worked. In Genie's opinion, science students were lucky. Learning by conducting experiments was much more interesting than merely reading books and listening to someone lecture.

Genie was so enthusiastic that she tried to practice her lab work at home. But not for long. Once, she brought home a big, dead rat to study its skeleton. To expose the bones, she boiled it in one of Grandma Yuriko's pots. Grandma would never have approved—but she wasn't home. Unfortunately, she arrived while the rat was still bubbling on the stove and before Genie could stop her, had lifted the pot's lid. Grandma's outraged scream rang through the apartment. Another time, the owner of a local pet shop gave Genie a small, dead monkey. She was looking forward to dissecting it later that day, and put it in the refrigerator so it wouldn't smell. But she forgot to alert the rest of the family. When Grandma opened the refrigerator, a horrifying figure stared at her from inside. "No more dead animals in this house!" she commanded.

During two of her college summers, Genie left the steaming concrete sidewalks of New York for the cool, green woods of northern Michigan to study at the University of Michigan's Biological Station. There she took field courses that allowed her to observe animals in their natural environment. Once again, Genie was learning through hands-on experience.

At the Biological Station, Genie shared a cabin with her childhood friend Norma, who was also studying zoology at Hunter. Both of them were on the college swim team and since the cabin was next to a lake, it was easy to keep in shape. Genie and Norma soon acquired other roommates—

Genie (right) with her best friend Norma, around 1940.

fish, snakes and a ground squirrel that Norma tamed. The squirrel not only shared their cabin, but also the generous supply of almond cookies that Nobusan sent from his restaurant in New York.

During her sophomore year, Genie made her first appearance in the press. Her unusual pet collection had attracted the attention of a Long Island newspaper. A reporter who came to interview her dared Genie to drape a large pet snake around her neck. When she accepted the challenge, someone snapped a dramatic photograph which appeared in the paper the next day. "Hunter girl starts new fad in necklaces," proclaimed the headline. Genie had to calm the irrational fears of many people, including the dean of Hunter College, who saw the photo. They assumed that all snakes are slimy and poisonous. Genie explained that a snake's skin feels dry and most snakes are harmless. Some, in fact, make excellent pets.

Years later she would have similar experiences helping people understand sharks and overcome their irrational fears about shark attacks. As she recently explained, "Very few people are ever attacked by sharks....Many millions of sharks are being killed by people every year. Sharks should be more afraid of us than we are of them."

2
THE MAKING OF AN ICHTHYOLOGIST

♦♦♦♦♦♦♦♦♦♦♦♦♦♦♦

By the time Genie graduated from Hunter College in 1942, America had entered World War II. On December 7, 1941, the Japanese carried out a surprise air attack on Pearl Harbor, a U.S. naval base in Hawaii. There were ninety-four American ships docked in the harbor, but Japan's primary targets were the eight American battleships. Three of the battleships were sunk or destroyed, as were eight of the other ships. The five remaining battleships were heavily damaged. And almost all the American planes were destroyed on the ground. The U. S. lost more than equipment—over 2,400 servicemen and civilians were killed or declared missing and almost 1,200 were wounded. The Japanese lost fewer than 100 men, only twenty-nine planes and five midget submarines. As a naval historian later wrote, "Never in modern history was a war begun with so smashing a victory by one side...."

The next day, the U. S. declared war on Japan, as well as its allies, Germany and Italy. Feelings against the Japanese were so hostile that over 120,000 Japanese-Americans, more than half of them women and children, were deprived of all their property by the U.S. government and isolated in remote prison camps. Despite this persecution, many soldiers of Japanese descent voluntarily joined the armed forces and fought bravely. Among them was a handsome pilot from Hawaii named Roy Umaki. He was Genie's new husband. They had met in New York and married just after her college graduation. Roy was sent to officers' training school and then went overseas, but wasn't allowed to fly planes because of his Japanese background. He and Genie spent most of the next few years apart.

In New York, Genie looked for work as a zoologist. There were lots of jobs in industries that supported the war, but very few jobs for inexperienced zoologists. After a long search, she finally took a full-time job as a chemist with a New Jersey company. Nonetheless, Genie was determined to become an ichthyologist, although it meant she would have to attend classes at night.

Columbia University was Genie's first choice for graduate school and she confidently sent in her application. The head of the zoology department was unenthusiastic about her future as a scientist. "Well, I guess we could take you but to be honest...you will probably...have a bunch of kids and never do anything in science after we have invested our time and money in you," he said. But the chairman of New York University's zoology department welcomed Genie. She immediately signed up for their graduate program in ichthyology, the first step toward achieving her goal.

Genie's friend, Norma, also enrolled at NYU and got a job with the same company. Both of them worked fifty hours a week, attended classes in the evenings and studied when they could—there was little time for sleep. It was hard not to doze off during a particularly boring *endocrinology* class, taught by a professor who lectured in a monotonous voice and showed slides in a darkened classroom. Even the perfectly typed notes they borrowed from a diligent, but confused, fellow student didn't help—all three of them failed the final exam in the course.

Genie had a very different reaction to the ichthyology course that was held at the American Museum of Natural History. It was taught by Dr. Charles Breder, head of the Department of Fishes. By coincidence, he had been the director of the New York Aquarium, Genie's favorite childhood haunt. His lectures were always exciting, and after class Genie could run down to the Museum's Hall of Fishes to look at the exhibits that related to what she had just learned.

Genie's course with Dr. Breder was the beginning of a close friendship, as well as a professional relationship. He became her first mentor and sponsored the research project for her master's thesis. Dr. Breder had long been interested in blowfish, fish that can puff themselves up like balloons in order to appear bigger to predators and scare them away. He asked Genie to investigate their puffing mechanism.

Blowfish belong to the *plectognaths*, a group of fish that would become one of Genie's research specialties. Plectognaths have only a few traits in common: they are slow swimmers with small gill openings; most of them live in tropical waters near coral reefs; and many of them are poisonous. They are remarkable mostly because they are so bizarre. For example, when frightened, one *species* stands on its head. Another grunts like a pig and sticks its head in a hole. And some blowfish have hundreds of spines that stick

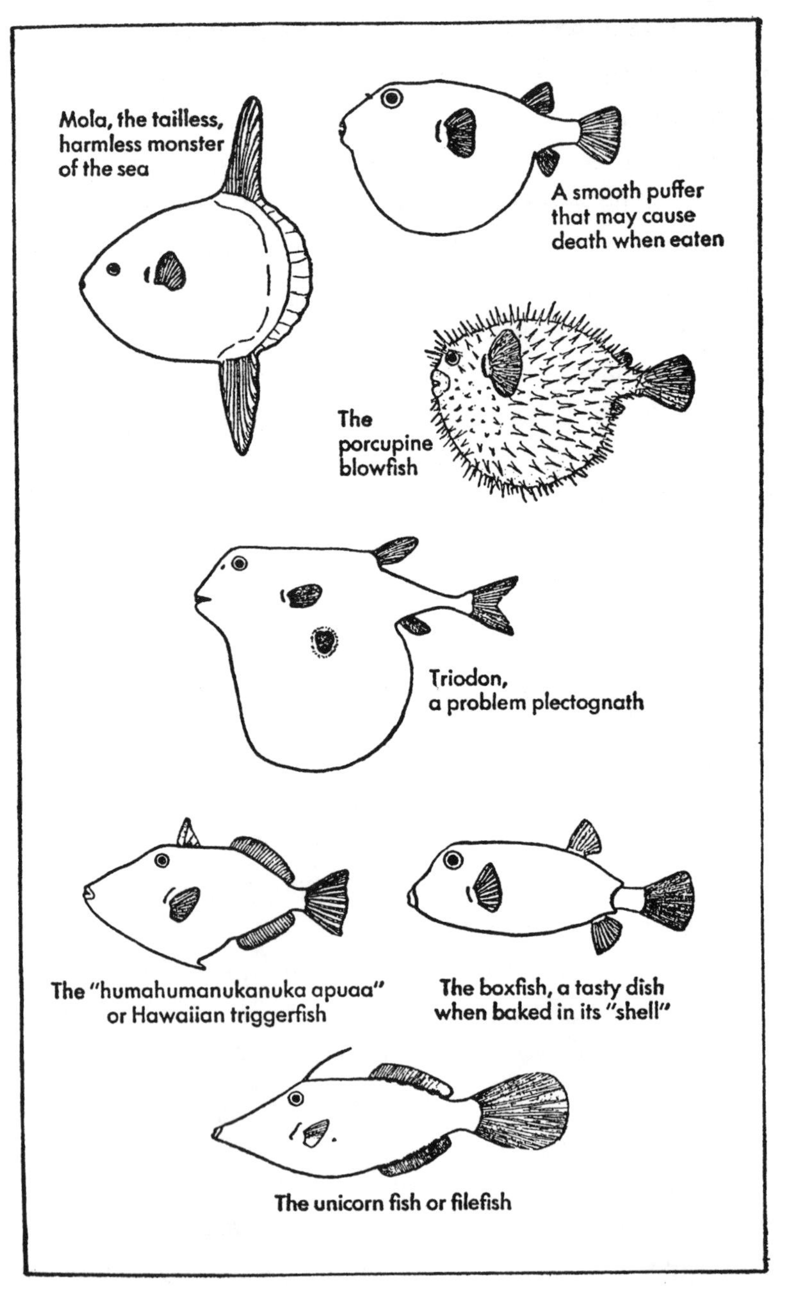

Genie's drawings of typical plectognaths give a very good sense of how odd these fish are.

out when the fish are frightened, making them look like angry porcupines.

Genie's research and drawings were so excellent that Dr. Breder published them along with his own work in the Museum's scientific magazine. She was thrilled. Later Genie joked, "...When he combined my final master's thesis with the publication of his own studies and my name appeared as co-author, my pride was as inflated as the blowfish."

Dr. Breder and her husband Roy weren't the only important men in Genie's life. In 1945, when her master's degree was almost complete, she attended the annual meeting of the American Society of Ichthyologists and Herpetologists. There she met Dr. Carl Hubbs, a world-renowned ichthyologist who was to become another mentor and close friend. He was so impressed by Genie's intelligence and dedication that he invited her to join him at the Scripps Institute of Oceanography in California as his part-time research assistant. He also offered her the chance to work toward a Ph.D. at the University of California in San Diego, where he was a professor. Genie eagerly accepted and was soon on her way to California.

She spent much of her time at Scripps doing what she loved best—learning about the ocean and its creatures. Dr. Hubbs, who also loved to swim, taught Genie how to dive using a face mask. Now she could observe fish living in the waters off the southern California coast in their natural environment.

Genie's first underwater dive was frightening. She saw a large greenish fish staring right at her and recognized it as a "dangerous" moray eel. Instantly, she popped up to the surface, followed by an amused Dr. Hubbs. He laughingly

The young zoologist on board a research ship belonging to the Scripps Institute, late 1940s.

explained that her glass face mask had magnified the eel's size and that moray eels attack rarely and only if bothered. Reassured, she dove back underwater to the exciting world of reefs and kelp beds.

Dr. Hubbs also taught Genie to dive wearing a metal helmet. It was connected by a hose to a compressed-air supply on board a ship. *Scuba* equipment, which has its own air supply, was not yet in general use. Although the helmet was heavy and awkward, it allowed her to walk on the bottom of the ocean. As a nine-year-old fascinated by the fish tanks in the New York Aquarium, Genie had imagined herself walking on the ocean bottom, meeting deep-sea creatures face-to-face. Her fantasy had now become reality.

Genie's first helmet dive was almost her last. She became so fascinated by the undersea world passing before her eyes that she didn't realize something was terribly wrong. A recently mended connection in her air hose had come loose, cutting off most of her oxygen. She almost blacked out before she managed to yank off the helmet and struggle to the surface, gasping for air. Dr. Hubbs and the other divers pulled her aboard the ship. After she rested for a short time, Dr. Hubbs insisted that she immediately dive again to overcome her fear.

A year later, in 1947, Genie learned that the United States Fish and Wildlife Service was looking for a scientist with skills in both chemistry and ichthyology to investigate the waters around the Philippine Islands. With her background, Genie was a good candidate. Here was a chance for her to use her scientific and diving skills in a faraway, exotic location—her idea of the perfect job. Dr. Hubbs encouraged her to try for this unique job. And she had no responsibilities to tie her down. She and Roy were spending most of their time far away from each other. Her mother had recently married their family friend, Nobusan. Genie was happy that her mother would not be lonely and was glad to have Nobusan in the family. She had always loved him like a father.

Genie was hired, but on her way to the Philippines she was delayed in Hawaii for several weeks. Even though the war was over, many Americans were still suspicious about people of Japanese heritage. She was told that the FBI wanted to check into her Japanese background, but never heard anything specific. It seemed that the problem was not her

TOP CAREERS FOR WOMEN IN THE 1950s

1. stenographers, typists and secretaries
2. other clerical workers
3. private household workers
4. saleswomen
5. elementary school teachers
6. bookkeepers
7. waitresses
8. miscellaneous unskilled workers
9. registered nurses
10. other service workers

ancestry, but her gender: A government official decided that it had been a mistake to hire a woman.

Genie was a pioneer in a man's world; there were only two other women ichthyologists in the United States, and neither of them was a diver. In the years following World War II, women ichthyologists were very rare. Genie had to compete for jobs in a profession that had traditionally accepted men. And as a Japanese-American, she had an added disadvantage. Genie realized that she would never be allowed to go to the Philippines and carry out her assignment, so she resigned. A male scientist replaced her.

During the weeks she spent in Hawaii, Genie fell in love with its beautiful scenery and spectacular tropical reefs. It was a fish-lover's paradise. One day, she went on an expedition with some other zoologists to track down a school of parasite-infested tuna that could threaten Hawaii's fishing industry. On the way back some of them went diving. As

Genie watched in amazement, one of the divers captured a huge parrot fish using a spear he had brought along. It was the first time she'd seen anyone spear a fish underwater. Genie vowed that one day she, too, would learn how to spearfish in order to catch specimens for her research.

She thought about staying in Hawaii, but reluctantly decided to return to New York and finish her doctorate. Professor Myron Gordon, a well-known fish geneticist at the American Museum of Natural History, hired her as his research assistant and agreed to supervise her Ph.D studies. Dr. Gordon specialized in platies and swordtails, the same kind of fish Genie and her mother had bought for their first home aquarium. In her doctoral *dissertation*, she answered some difficult questions about their reproductive behavior.

In the laboratory, it's easy to get the two different species of fish to mate and have *hybrid* babies. But in the Central American rivers where platies and swordtails live, scientists have never found a hybrid even though the two species often live side by side. Genie spent many hours observing their mating behavior in Dr. Gordon's lab. She developed a technique for producing "test-tube" babies in female fish, and used it in her experiments. The results of her research explained why platies and swordtails never have babies together in the wild.

Genie spent three years working in Dr. Gordon's lab at the Museum. As she had in college, she escaped to the outdoors during the summers. At Woods Hole, Massachusetts and in the Caribbean she went diving and observed the local fish. She noticed a strange Caribbean plectognath, a kind of filefish that had the odd habit of standing on its head. Genie's high school hero, Dr. William Beebe, had observed

HOW TO MAKE A "TEST-TUBE" FISH

To make her fish pregnant, Genie used a common piece of lab equipment called a pipette. A pipette looks something like a glass straw with a long narrow tip. Like a straw, a pipette is used to draw up liquids. Genie used tiny pipettes to remove packets of sperm from male fish. She injected the packets into the reproductive organs of female fish, who had never been with males before. There the sperm could fertilize the female's eggs. When the first twelve fish babies were born, everyone in the Museum congratulated Genie. No one in the United States had ever been able to produce a "test-tube" fish before.

Genie making a test-tube fish at the American Museum of Natural History, 1947

this, too. He believed it was *camouflage behavior*, since the species of filefish he studied are long and slender and look like seaweed when they stand on their heads. But Genie's filefish are short and plump; through careful observation she learned that the males use headstanding as a way to establish who's boss. Whenever two males meet, each one stands on his head, spreads his fins, sticks out his "file," and vibrates his body. The smaller fish usually gives up, folds his fins and backs away from the larger one. The male who causes all the others to back away is the top fish. These contests continue until all the males find their place in the group. The weakest male may starve to death because the others eat all the food.

By the time she was twenty-five years old, Genie was already a skilled scientist and had begun to make her own discoveries. In 1949, she heard that the Office of Naval Research needed qualified scientists for a four-month project in the South Seas. As a result of the war, hundreds of tiny South Sea islands became U. S. possessions. The Navy wanted to find out whether the waters surrounding them could be used for commercial fishing and it was important to learn which fish were poisonous to eat. The Navy was especially interested in blowfish, whose anatomy and behavior had been Genie's special research project when she worked with Dr. Breder. But there were two major obstacles: The Navy didn't usually send a lone woman on a rugged expedition to collect fish in strange places; and, the people of the South Sea islands believed a woman would bring bad luck on fishing expeditions. To Genie's surprise, she was hired.

3
SOUTH SEAS ADVENTURES

Genie had only a few weeks to get ready for her trip to the South Seas. She completed the last experiments for her doctoral dissertation and hurriedly packed clothes and equipment. On June 17, 1949, she boarded a military plane headed for Hawaii. She could hardly believe that she would soon be working in the romantic South Sea islands she had read about as a child. The islands, officially known today as Micronesia, are located in the Pacific Ocean just north of the equator. Micronesia is made up of six hundred tiny islands which stretch from New Guinea, a large island off the north coast of Australia, up toward Japan.

Fifty years ago, very few people visited the South Sea islands for any reason. And it was even more unusual for a woman to travel by herself to such a remote location. The trip took several days. There were no commercial jets, only propeller planes that flew much more slowly and had to stop for refueling every few hours. Nowadays it's common for tourists to fly nonstop from California to Australia in less than twenty-four hours.

Genie had fun during her short layover in Hawaii. She visited friends, did some last-minute shopping and picked up her Navy I.D. card and some photographic equipment. By the time she boarded a military plane to the South Seas, her baggage weighed in at 260 pounds. The plane developed engine trouble on the way to her first official stop, the island of Guam. It was forced to turn back to Kwajalein, a tiny island with a large naval base.

Genie had to wait two days for space on another plane, but she made the most of the unexpected opportunity. She decided to do some collecting in the *tide pools* on the island. In order to learn about the anatomy and physiology of fish, ichthyologists must dissect specimens and study them under a microscope. The Navy was happy to provide Genie with volunteers—all male, of course—and to lend her the unused morgue on the base for a lab.

Genie tried out a liquid chemical called rotenone she had brought along for use in tide pools. Spreading rotenone in a tide pool is the best way to collect many local species of fish at once. Rotenone is made from the roots of a special group of plants. Native fishermen in tropical countries around the world have used these plants to catch fish for hundreds of years. The fish become stunned and float to the surface because rotenone prevents their gills from absorbing oxygen. But the fish are safe to eat. Rotenone doesn't harm or kill plants or air-breathing animals, including humans, who accidentally swallow the water.

Often the most interesting fish in a tide pool are almost impossible to find. Many of them, like the goby, are less than an inch long when full-grown. Others are well camouflaged, like the seahorse which hangs by its tail and mimics a "leaf" of seaweed. And still others hide in holes and crevices. After adding a small amount of rotenone to a tide pool, Genie and her volunteers were able to use dip nets to collect hundreds of fish that floated to the surface. It was a new experience for the men, who were excited about the fish they caught. They netted a large brown sea bass with brilliant blue spots that was supposed to be poisonous to eat; a rare trumpet fish with a long skinny body and a matching snout; a striped damsel fish called a Sergeant Major; and a two-inch-long, venomous scorpion fish with needle-like spines on its back fin. At the morgue, the men helped Genie sort her specimens and preserve them for shipment to the United States. Her work as a Navy scientist had begun.

The next day Genie flew to Guam, an island that had been a United States possession since 1898. The base commander had no trouble deciding where she should stay—she was given a large room in a building used for visiting Navy wives and children. But there was some confusion about where she should eat her meals. Finally, a small, separate dining room was set up for Genie in the same building with the male officers. She was even assigned two waiters, who took great pride in keeping her glass and plate full. At first the arrangement seemed strange, but she grew to like having her own dining room.

Once again, Genie collected fish in tide pools. In seven outings, she gathered hundreds of specimens, many of them different from those she had found on Kwajalein. She also found some interesting fish caught by a local fisherman. Though his catch was small, it included seven poisonous puffers, exactly the type of fish the Navy had hired her to study. After much hesitation, because he was afraid she

POISONOUS AND VENOMOUS FISH

It's difficult to figure out whether a particular kind of fish is poisonous to eat because one that is poisonous to eat in one location may be perfectly safe to eat in another. Many fish that are poisonous to eat are found in tropical waters near coral reefs. Some fish are poisonous only after they grow to a large size; as they grow they eat other poisonous fish and the poison stays in their bodies. Other fish are poisonous only at certain seasons. In some fish, the poison is everywhere; in others, only in certain organs. The poison does not break down at high temperatures so cooking the fish doesn't remove the danger. Eating a poisonous fish can cause serious illness and even death.

Fish that are venomous to touch are always venomous without regard to size, season or location. There aren't many kinds of them and they are easy to recognize. The poison in a venomous fish is concentrated in certain parts of its body. If those parts are removed, the fish is safe to handle and eat. For example, being pricked by the venomous spines of a scorpion fish causes an extremely painful sting; but once the spines and poison glands are removed, the scorpion fish can be eaten.

would eat them and get sick, the man finally let her have the fish. Genie prepared samples and sent them back to San Francisco for a poison analysis.

Before Genie left Guam, she spent a day fishing with Ramon Quenga, the son of one of the island's best fishermen. Ramon's uncle came along, too, but sat silently in the boat and watched. The Quengas' traps were full of all kinds of fish and shellfish, including the plectognaths that were Genie's special interest. At lunch time, Ramon's uncle played a practical joke on her. He picked up a still-live squid, cleaned it and took a bite. Then he offered it to Genie. She was always ready to try native delicacies, but it was hard to ignore the squid's staring eyes. When she finally tried a bite, the uncle began to laugh wildly and Ramon looked embarrassed. Genie's daring impressed Ramon's uncle, and from then on, he treated her as a friend.

When they returned from their fishing expedition, Ramon's father invited Genie to supper. Mr. Quenga proudly showed her around his house, which had dirt floors and an open hearth for cooking but also a modern refrigerator and washing machine. After a delicious feast, Genie returned to the Naval base carrying gifts of food, all the plectognaths from the day's catch and a perfect *nautilus* shell given to her by Mr. Quenga.

Next, Genie flew to Saipan, a small island less than an hour away from Guam. She stayed with Commander Sheffield, the governor of the Mariana Islands, and his family. The Sheffields went on some of her collecting trips and helped sort fish in the evenings. They were a fun-loving family and included Genie in their picnics, sunset barbecues and moonlight swims.

Saipan's tide pools were the best collecting sites that Genie found in the Pacific. They were crowded with coral colonies that were home to thousands of fish. One pool contained fifty-seven different kinds. Genie caught a brown-and-white moray eel, edible in the Philippines but deadly in Saipan; a comical little boxfish, bright orange with large black polka dots, huge dark eyes and a tiny, round mouth; and a tiny, emerald-green filefish with golden spots, a bright yellow mouth and blue-and-orange eyes. Genie was sorry to

Siakong preparing to spear a fish. Genie called him the best diver in the world.

leave Saipan and the Sheffields, but she had many other islands to explore.

From Saipan, she flew on to Koror, one of the Palauan Islands. There she met the most colorful character of her entire trip, a fifty-year-old fisherman named Siakong. Genie described him as a "betel-chewing, wife-beating drunkard." In his shabby everyday clothes, he looked like a bum; but when he wore his diving outfit—a red loincloth and a pair of homemade goggles—he looked like a prize-winning body builder. Siakong was able to stay underwater on one breath for an unbelievably long time and was an expert with throw nets and spears. He could even catch fish with his bare hands!

To catch the smaller fish that lived in tide pools, Genie was able to use fish poison and nets. For larger fish in the open ocean, she had to learn different techniques. She couldn't helmet dive since there was no compressed air available on the islands. Instead she had to dive with only a mask. Siakong taught her how to stay down on one breath for a much longer time. He also taught her to spearfish, something no woman of the islands was allowed to do. Together, they made a colorful pair as they explored the nearby coral reefs.

On one of their dives together, Genie unknowingly rested her arm on the open shell of a giant clam. It was so well camouflaged that it looked just like part of the coral reef. Siakong warned Genie just before the clam began to close its shells. Both of them had heard terrifying stories about giant clams that trapped and drowned divers by clamping down tightly on an arm or leg. But giant clam meat, especially the muscle that holds the shell closed, is delicious. Siakong was an expert at handling the huge clams and knew how to remove their tender meat after pulling one up into the boat. He and Genie often enjoyed a picnic lunch of raw clam meat on their outings.

One day, Siakong and Genie found an enormous clam—it was almost four feet across and weighed about five hundred pounds. The clam was on the bottom of the ocean, so far down that Genie couldn't hold her breath long enough to reach it. But Siakong could and did. When he didn't return to the boat, Genie became alarmed. She went down to have a look and froze in horror. Siakong's arm seemed to be caught between the clam's shells. She was sure he would die and quickly surfaced to get help. The man who piloted their boat didn't seem to understand. In a panic, she prepared to dive again. Suddenly, Siakong appeared beside her laughing and holding out the biggest clam muscle she had ever seen. He had never been in danger but realized that Genie would think he was. It was his idea of a clever practical joke.

From her base in Koror, Genie was able to explore several more islands. Wherever she traveled, she collected specimens from tide pools and went spearfishing near coral reef formations in the open ocean. She added many more plectognaths to her research collection, including an unusual filefish that swam with and mimicked a poisonous pufferfish. She also had several startling encounters with barracudas and sharks.

As she visited different islands, Genie learned a lot about their human, as well as their fishy, inhabitants. She couldn't speak Palauan and had to communicate in a mixture of Japanese and English. But she was so friendly and interested in how the people lived that she was quickly accepted. On one of her trips to an island near Koror, she was housed for the night in the men's clubhouse, called an *abai*. Normally, women were not permitted inside the *abai*, but Genie was considered such a distinguished visitor that an exception was made. Her roommates for the night were six men, a situation which concerned the chief of the village so much that he sent a chaperone for Genie—another man!

She spent a week on the island of Kayangel, as a guest of Siakong's sister, Uredekl. Like her hostess, Genie slept on a mat on the floor. There were no plates or forks and they ate with their fingers, using large leaves as plates. At first, Uredekl was shy and didn't say much, but after a day or two, she and Genie became good friends. After spearfishing on the reefs all day, Genie spent her evenings getting to know the

Outside a ceremonial hut with the high chiefs of the South Seas island, Fais.

women of Kayangel and listening to their stories, which Uredekl translated into Japanese. At the end of the week, she was sorry to see the boat arrive to take her back to Koror.

During Genie's last few weeks in Micronesia, the Navy allowed her to join two field trips to some of the more remote islands. She collected several interesting plectognaths, but especially enjoyed the chance to meet the native people. On the island of Fais, all of the high chiefs turned out to greet the Navy expedition. According to Genie, "Their faces were serious and dignified, and they presented a weirdly beautiful picture. Their bodies were heavily tattooed from the neck down with vertical stripes and designs of various types, some of which I made out to be fish." The women of Fais wore nothing but skirts of hand-woven cloth. They adorned themselves with bracelets, beaded necklaces and belts and had tattoos all over their arms and legs. At night, the women danced before a bonfire for Genie and the other guests. The older women were the best dancers, but Genie observed, "the young girls made up in attractiveness what they lacked in dance training."

Genie's last stop was Ulithi, a group of forty-nine islands, only seven of which were inhabited. On the island of Mog Mog, she met King Ueg, the supreme ruler of all Ulithi. The king had been paralyzed by polio and traveled around the island in a small wagon pulled by one of his subjects. Although his legs were withered, his face was handsome and dignified. He had blue eyes, lots of curly black hair and a warm smile. His only ornaments were his tattoos and a simple comb in his hair. Genie's description of their meeting is unforgettable: "I had never dreamed that at my first meeting with a king, I would be in bare feet, bathing suit, and pigtails, and the king in a G-string!"

Genie would have loved to continue her adventurous life in the South Seas, but she completed her fish survey for the Navy project soon after her visit to Ulithi. In the fall of 1949, she returned to New York to finish the work on her doctorate. She kept some reminders of her stay on the beautiful islands of Micronesia in her laboratory at the American Museum of Natural History. The jars of preserved fish may have looked dull to others, but for Genie they brought back memories of brilliantly colored fish swimming among coral reefs in the clear, warm waters of the South Seas.

4
THE LURE OF THE RED SEA

♦♦♦♦♦♦♦♦♦♦♦♦♦♦♦♦♦♦♦♦♦♦♦

Several months later, Genie read an article written by Dr. Hamed Gohar Bey, the director of a marine laboratory on the Egyptian coast of the Red Sea. The article was about local clown fish that looked just like clown fish she'd seen in Micronesia. This wasn't the first time Genie had noticed a similarity between fish in the two locations. The plectognaths she'd studied in Micronesia also seemed similar to the plectognaths she read about in the Red Sea. Genie realized that if she could get to Egypt, she would have a unique opportunity to compare fish in the Red Sea with fish in the South Seas. And the timing was right, since she'd finally completed all the requirements for her Ph.D. All she needed was a sponsor to provide the funds. Genie applied for—and won—a Fulbright Scholarship that paid her expenses for a year studying plectognaths and other poisonous fish in the Red Sea.

But Genie's decision was complicated by events in her personal life. She and Roy had been divorced in 1949. Soon after, she fell in love with Ilias Konstantinou, a young Greek doctor who was working at a hospital in New York City. Both of them were busy with their careers. Sometimes when they went out on dates, Genie ended up spending the entire time waiting in the hospital lobby while Ilias operated on a patient. On other dates, Ilias helped Genie to translate ichthyology articles from German into English or figure out the results of her experiments. He thought her trip to the Red Sea sounded exciting and encouraged her to go. They were married just before she left, alone.

Once again, Genie quickly packed some belongings and left New York City to explore a faraway place. Although the Middle East was somewhat closer than the South Seas, the trip was also long and difficult. And very few Americans made the journey.

One of the most famous stories in the Bible is in the Book of Exodus. It describes how the waters of the Red Sea parted, allowing Moses and the Israelites to escape from Egypt. The Red Sea is a long, narrow arm of the Indian Ocean that separates Africa from the Arabian Peninsula. It is about the size of California. On the north it forks into two prongs—the Gulf of Aqaba on the east, the Gulf of Suez on the west. Between the two gulfs lies the Sinai Peninsula. For centuries the Red Sea was one of the world's greatest trade routes. Nations along both of its banks fought to control it. And when the Suez Canal was built in 1869 to connect it to the Mediterranean, the Red Sea became more important than ever.

The Red Sea is unique. Its waters are among the warmest and saltiest in the world. A trench 1.5 miles deep slices up its middle. In certain places near its shore, the sea floor suddenly drops thousands of feet. The Red Sea is home to a vast assortment of marine life. About 20 percent of its species live nowhere else on Earth because the narrow openings at each end of the sea make it difficult to enter and leave. Huge coral reefs fringe its shores. They are made of all kinds of corals that grow in colorful clusters. Some are odd-shaped; others grow in familiar shapes like branches or fans. For a long time, Red Sea life had been almost entirely ignored. But that was about to change. Genie would be the first ichthyologist to study Red Sea fish in seventy years.

On Christmas Eve, 1950, she arrived in Cairo. Two weeks later she reached the Marine Biological Station at Ghardaqa, Egypt. It was part of Fouad University in Cairo where Dr. Gohar was a professor. Genie was the first woman to work there. The Station was a small group of buildings on the eastern edge of the desert, four miles from the village of Ghardaqa. The buildings, surrounded by huge empty stretches of sand, looked out on the clear blue waters of the Red Sea. The Station consisted of a library, museum, an office building and laboratories. There were also cottages for the small staff and a cottage for visiting scientists.

Genie described the Red Sea as the most extraordinary place on Earth. Its name comes from the tiny red algae that live on its surface and make the water look red during certain seasons. But Genie says the sea should be named for the glowing pink light reflected into it from the towering desert cliffs at sunrise and sunset.

From the back balcony of her cottage, Genie could look over the sea to the mountains in the distance; from her front porch she could see the desert and high cliffs beyond. Nearby stood a pink-and-white striped building that was the mosque, used by Muslims to pray. She saw them wash themselves in the sea before entering, and heard them chanting passages from the *Koran*, their holy book. Sometimes in the evenings, she could hear the sounds of laughter and singing as her neighbors danced to the music of homemade drums and simple flutes.

The fishermen, sailors and divers who worked at the station came from the Nile Valley, the Sudan and from the area around Ghardaqa. They spoke only Arabic and couldn't read or write. It was very different from the French and Japanese Genie already knew. The men listened patiently as she fumbled for words, kindly taught her new ones and politely pretended to understand when she spoke. But the children laughed at her mistakes—they didn't know how to pretend.

The workers and their families all lived nearby. During the first few weeks, Genie rarely saw any women. When she did, they were covered from head to foot in long robes and heavy veils, while she wore a bathing suit most of the day. In the South Seas, where women wore only grass skirts, no one had been upset about her skimpy clothing. They said she was wearing too much on top and not enough on the bottom. In the Egyptian desert, however, she was a strange and shocking sight. But gradually the women got used to seeing Genie and became more friendly. They invited her into their homes where they could unveil their faces and talk and laugh together.

Genie sometimes had trouble communicating with Mohammed, the Station cook, because he spoke only Arabic. He was a wonderful cook and could prepare all kinds of fish, in all kinds of ways. He cooked octopuses, sea turtles, dugongs, giant clams—anything that the fishermen caught.

Genie taught him how to cook Japanese-style when Nobusan sent a special package of Japanese seasonings. But Mohammed was never satisfied serving raw fish—he always wanted to do more than just slice and arrange it on a plate. Once, in her halting Arabic, Genie asked Mohammed to prepare a special Japanese meal in honor of Dr. Gohar. She wanted the table to be set with her beautiful new ivory chopsticks. When everyone sat down to eat, Mohammed proudly brought out a plate holding the first course—Genie's chopsticks.

The Station had a pier leading out to the sea. At its end were the small laboratories with running sea water and many large aquariums for keeping live specimens. Manta rays, nurse sharks, sawfish, guitar fish and a hundred other fantastic marine animals swam back and forth. As a child in New York City, Genie could only dream about swimming with sharks and exotic fish. At Ghardaqa she could simply walk outside and dive into the water to swim with them.

To catch certain kinds of fish, Genie used rotenone as she had in the South Seas. But she had to trap most specimens instead, because there weren't many good tide pools and she had a limited supply of the chemical. With one or two sailors, she would paddle a dugout canoe to one of the huge coral reefs near shore. So many thousands of small fish, crabs, marine worms, sea slugs and other creatures lived there that they could spend an entire morning collecting in one limited area.

Other small specimens were caught using a heavy net. It was difficult work for the sailors because the boat had to be kept at a slow steady speed. Then the heavy net had to be hauled up and emptied very carefully so the specimens wouldn't be damaged. The sailors were often stung, cut and clawed as they sorted out Genie's "treasures" from the piles of debris that also got snagged in the net.

Sea horses, and their relatives the pipefish, were often tangled in the trawl nets. Genie thought they didn't look at all like fish. When they swam, which wasn't often, it was difficult to tell what made them move because their tiny twirling dorsal fins were almost invisible. What interested her most was that both sea horses and pipefish are the only species in which the male becomes pregnant and gives birth. The female deposits her eggs in the male; the male then fertilizes them. A pouch on his belly swells as the babies grow. The female visits every morning during the pregnancy. Twining their tails, the parents sway together in a sort of dance. After a few weeks, the male gives birth to as many as 200 tiny babies who must survive on their own in the danger-filled ocean. Almost immediately, he gets pregnant again.

Genie also studied other fish, like parrotfish, razorfish and moray eels, that lived around the coral reefs. She and a few fishermen set out in a sailboat, pulling a dugout canoe filled with 60-foot-long nets and other equipment. They looked for a shallow area as they approached the reef; then quietly dropped anchor, changed boats and paddled to the reef's edge. Silently they slipped into the water and waded over to the reef, holding the huge net. When the net was spread into a semi-circle around the fish, Genie and the men "suddenly burst into a wild dance over the reef, shouting, splashing, throwing rocks, and hitting sticks in the water." The fish were scared and tried to escape; but they were trapped. Sometimes over a hundred were caught. The fishermen liked the huge numbers but Genie cared about the many varieties.

Gomah and Atiyah were Genie's spearfishermen. But even at their best, they couldn't compare with Siakong and the other spearfishermen of Micronesia. Gomah could dive no deeper than fifteen feet; and Atiyah was so nearsighted that he often missed the fish Genie wanted. Neither of them liked getting their heads wet. Instead of diving, they preferred to walk along the edge of a reef in chest-high water, holding a spear in one hand and a glass-bottomed box in the other. They peered through the glass into the water to search for fish. Unlike the natives of Micronesia, the fishermen at Ghardaqa didn't have goggles. But after Genie insisted that Gomah try wearing a face mask, he discovered how much better he could see underwater.

The Red Sea fishermen used spears with handles made of metal. They were much heavier than the wooden-handled spears of the South Seas, and clumsier to use. It could also be dangerous to use a metal spear. Once, while spearfishing, Genie saw a large flat fish lying still in the sand and thought it might be a poisonous sting ray. Just as she was about to spear it, she realized that it wasn't a sting ray, but an electric ray. If the spear had entered the electric ray, the metal handle could have conducted an electric charge from the fish to Genie, giving her an unpleasant jolt.

Genie's days were busy; she never felt lonely or homesick. She spent most mornings collecting fish in the sea. As soon as she returned, she separated the living specimens into groups that could get along together and put them into aquariums. Then she quickly jotted down descriptions of the dead fish, before they were preserved and their colors faded completely. They could be dissected and studied later. When she discovered new or rare species, she sketched them carefully and described their anatomy.

The rest of the day Genie studied her catch and recorded her observations. In her journal Genie wrote lovingly, often poetically, about the sea creatures. She described a type of sea slug called a *nudibranch*: "A beautiful, rosy-red mass of jelly with a ring of delicate white feathers on one end. If you pick it up and stroke the soft mass in your hands, it opens up like the circular skirt of a dancer. And then if you put it back in the water—the skirt begins to dance!...it swims through the water as if worn by an invisible ballerina."

After her lab work, Genie would often go to the library or visit the museum. The museum had an excellent collection of Red Sea animals and plants. Eventually she and Dr. Gohar redid the fish exhibit and with the help of assistants, labeled everything in English and Arabic.

Sometimes Genie went to Cairo to give lectures and shop for souvenirs. She also left Ghardaqa for some short vacation trips, visiting other cities or traveling over hot desert sands and through valleys between looming mountain peaks. Almost no one lived in the mountains anymore, but they had an interesting history. Centuries before, they had been inhabited by thousands of people. Genie visited the remains of ancient Roman cities, built for traders and workers who mined gold and cut granite from the mountains' sides.

Visitors often arrived at the Station, some staying for a few days, others a few weeks. The American ambassador, some Egyptian princes and many university professors came to learn and share ideas. In June 1951, Genie had a special visitor—her husband Ilias. He spent several weeks at Ghardaqa where they shared a honeymoon that had been delayed for six months.

He and Genie spent a lot of their time together in the water. They swam in a pool outside the lab that housed nurse sharks. Most often they swam in the sea. Ilias was surprised to learn about the many harmless shark species. Genie also taught him about the more dangerous varieties, such as the tiger, mako and hammerhead. It was important to understand the perils of the sea, as well as its pleasures. Ilias became an enthusiastic spearfisherman, learning to launch the metal spears with a speargun. Genie had been using the lighter, wooden-handled spears she preferred, but learned the speargun technique, too.

Ilias's inexperience could have gotten him into trouble. One day, he was diving on one side of some corals while Genie was chasing a fish on the other. Suddenly she saw an enormous barracuda. Barracuda are long, slender, silvery fish whose mouths are filled with large, sharp teeth. They are ferocious and may strike at anything that gleams. She was afraid the barracuda would attack, but it didn't seem to notice her. Taking care not to startle it, she quietly swam away. But Ilias was kicking and splashing among the corals. Genie was afraid the barracuda would mistake the bright white sneakers on his feet for two little fish. She tried to get his attention without alarming him, but Ilias didn't understand. He thought she needed their boat. Splashing water all over, he quickly swam to get it. Genie waited anxiously as he lifted himself in and paddled to where she was standing. Breathlessly, she told Ilias about the danger he had escaped. But he was disappointed that the danger had been "just" a barracuda, not a shark.

By the time she left for home, Genie had collected three hundred species. Some of them were very rare; three of them were unknown. Of all the many species, only about thirty were venomous to touch. They were mostly scorpion fish and sting rays. And, like the South Seas, the only poisonous-to-eat fish seemed to be puffers.

Genie's year in Egypt was over. Her travels had taken her to places most people could only imagine. She returned to the United States and joined Ilias in Buffalo, New York, where he was an intern in orthopedic surgery. She spent her time organizing the hundreds of pages of notes and sketches she had made in order to write several scientific papers. She also began to write her first book, an autobiography called *Lady with a Spear*. The book was published in 1953, when she was only thirty years old, and became a bestseller. It was filled with fascinating stories about her adventures with the fish and people of Micronesia and the Red Sea.

5
AN UNEXPECTED SUCCESS

♦♦♦♦♦♦♦♦♦♦♦♦♦♦♦♦♦♦♦♦♦♦

In 1952, the year after Genie returned from the Red Sea, she not only wrote her first book but also gave birth to her first child—a daughter named Hera. Shortly after, the family moved to New York City. Ilias finished his medical training and Genie became a professor in the biology department at Hunter College. She also worked as a researcher at the American Museum of Natural History and lectured at high schools and universities around the United States.

In 1954, Genie accepted an invitation from Anne and Bill Vanderbilt to give a talk in Englewood, a small town on Florida's west coast. The Vanderbilts lived near the water and their ten-year-old son spent a lot of time exploring the seashore. His room was lined with aquariums filled with all sorts of strange sea life that he brought home. The Vanderbilts were fascinated by his hobby, but couldn't identify most of the fish even with the help of the local fishermen. And there were no experts in the area to ask.

Anne happened to read *Lady with a Spear* and persuaded Bill to read it, too. They were very impressed and wanted to meet Genie in person, so they invited her to give a lecture at the local public school about her experiences. The turnout was large. Genie was surprised that so many people were interested in Red Sea fish. From their comments, she learned that many of the unusual fish she discussed had been seen on Florida's west coast. After her talk, the Vanderbilts invited Genie to set up a laboratory in the area to study the local marine life. Bill and his brother Alfred owned a huge piece of land on the Cape Haze peninsula, a perfect location on the Gulf of Mexico. The two of them donated the money, as well as the land, for the lab. Genie would be its founder and director.

The timing was right for the Konstantinou family. Genie was pregnant with their second child. She and Ilias preferred to raise the children in a Florida beach house, rather than a New York apartment. He also liked the idea of starting a practice in Florida. In January 1955, just six months after her first visit, Genie, Ilias, Hera and one-month-old Aya moved to Florida.

Genie opened the Cape Haze Marine Laboratory as soon as she found a babysitter for her daughters. Beryl Chadwick helped open the Lab and became her assistant. He was an expert fisherman and knew a lot about the local fish. The Lab was in a small wooden building. Its one room was about the size of a small living room, with sinks and shelves to hold specimens. Nearby was a small dock for a 21-foot boat donated by Alfred. Genie's goal was to study the fish and plant life of west coast Florida, but she and the Lab soon became known for an activity no one had predicted—the study of living sharks.

Right after opening day, Genie got a call from Dr. John Heller, Director of the New England Institute of Medicine.

He needed fresh shark livers for his research on cancer. Dr. Heller had been hunting for sharks in the Caribbean, but after many weeks he hadn't found any. Since Genie had always been interested in sharks, she agreed to help. Beryl rigged a thick, 300-foot-long rope, with 16 evenly spaced steel shark hooks, securely attached with rope and 3 feet of steel chain. Within a week, they caught 12 adult sharks.

Most of the sharks were dead or dying by the time Genie or the others found them. If they were alive, water got into their stomachs while they were towed behind the boat back to the Lab, and they died. If a shark managed to survive, there was no place to keep it.

Very little was known about living sharks. Although she learned a lot about their anatomy by dissecting dead sharks, Genie realized that in order to study their behavior and really understand them, it was important to keep them alive and watch them in their natural environment. She had some workers build a large pen, enclosed by strong posts, next to the dock. She began to spend countless hours watching the sharks and other large fish in the pen and the ocean and kept records of her observations. Often Genie swam with them in order to observe them more closely.

There were eighteen shark species in the waters off the Florida coast. The local fishermen who caught them gladly gave them to the Lab. Most of the sharks she studied lived within 8 miles of the shore; they were not the deep-sea varieties. Genie had her choice of hammerheads, blackfins, dogfish, nurse, tiger, bull, lemon and sandbar sharks—most of them 5 to 11 feet long.

Dr. Perry Gilbert, a Cornell University professor who specialized in shark research, visited the Lab. He invented a technique that helped Genie's research with live sharks. A special chemical was sprayed into their mouths and over their gills. The fish became unconscious almost immediately and stayed that way for at least ten minutes. While unconscious, they could be brought back to the Lab without a struggle.

Genie began writing reports about her shark research and her success at keeping sharks alive in captivity. Her discoveries attracted scientists from around the world. Newspapers published articles about her work and her previous adventures. As the publicity grew, Genie earned the name that she became known by—the "shark lady."

During its first year, many scientists came to study and work at the Lab. In the summer, crowds of schoolchildren visited to watch the fish and learn about them. People living in the area often contributed animals that they caught or no longer wanted as pets. An alligator, snakes, turtles and birds found a home there and at times it seemed more like a zoo. The Lab overflowed with people and sea life; clearly, it needed more space.

Research grants and financial help began coming in from organizations like the National Science Foundation and the Office of Naval Research. By the end of the year, the Lab had expanded to four rooms. The original lab, lined with shelves, was filled with preserved fish, invertebrates and preserved sea plants. All were catalogued, labeled and arranged. A central table held the microscopes. An aquarium room was added. It had over thirty aquariums in all sizes, plus large holding tanks for sorting and studying live fish, and a pump that kept the tanks supplied with fresh sea water. There was also a small office and a library room for books and scientific journals. More shark pens were built for the Lab's increasing shark population. Eventually a museum was created for exhibiting collections of the local marine life.

Sometimes visitors ignored the "Danger—Sharks" signs posted around the shark pens. One afternoon, Genie glanced up from her work and saw a little boy sitting on the feeding platform of one of the pens. His feet were dangling over the water, while a shark swam silently below. Genie dashed out of the Lab and pulled the boy to safety. His parents had paid no attention to the danger signs. Since they hadn't seen any sharks in the water, they'd assumed the pen was empty. Genie knew the shark had already eaten; otherwise, it might have tried to grab the boy's feet, mistaking them for fish.

Sometimes people sneaked into the pens when the Lab was closed. Once, Genie saw a man and three young women in a boat. They pulled up alongside a pen that held twenty newborn sharks that were being carefully nurtured. The man caught one of the two-foot baby sharks, twirled it around his head and threw it across the pen. When he saw Genie rushing toward them, he sped away. It was sad that although people were terrified of shark attacks, they weren't concerned about the effects of people attacks on sharks.

In the twelve years that Genie was its director, the Cape Haze Marine Laboratory became one of the most respected institutions for marine research in the world. It was particularly famous as a center for shark studies. Over the years, Genie and her team handled thousands of sharks and other marine animals. Hundreds of scientists came from around the world to work with Genie; many of them returned again and again. Sometimes they arrived with their families. Children were encouraged to get involved in many of the

Measuring a tiger shark at Cape Haze.

Lab's activities. They helped haul up sharks to be weighed, fished from the dock, studied specimens through microscopes. Sometimes, when Genie dissected sharks on the dock, she gave the children jobs to do—measuring parts of the intestines, washing a shark stomach, or hosing off the dock when she was done. The staff organized summer programs for students from all over the United States. Some of them stayed for several weeks, working as volunteer assistants.

Although sharks were the focus of the Lab's research, they weren't the only fish Genie studied. A few months after arriving in Florida, she began diving again. Sometimes she wore only a face mask, but for deeper dives she used scuba equipment. Genie explored the murky green waters of the Gulf of Mexico. They were very different from the clear waters of the Pacific and the Red Sea that were filled with beautiful coral reefs. The reefs off of Florida's coast were made mostly of rock. But, like the coral reefs, they were home to many kinds of fish.

She noticed hundreds of tiny grouper fish, called *Serranus*. Although these fish looked ordinary, Genie noticed something very puzzling about them. All of them seemed to be females and all of them were pregnant, their bellies swollen with eggs. But where were the males? After months of study Genie found the answer: Each *Serranus* has both male and female organs. A few other vertebrates can change sex, too. But *Serranus* are unusual because when they mate, they can switch sex roles in just ten seconds, faster than any other fish. And what makes them unique is that, unlike any other animal, they can be both sexes at once. A *Serranus* can fertilize its own eggs when it can't find a mate. Genie had made an important discovery—an ordinary-looking fish that was extraordinary, a single fish that could be both mother and father to its offspring.

Genie usually went diving two or three days a week. On other days she and Beryl, and sometimes others, took field trips in the boat. They looked at the wildlife, collected specimens and often found something good to eat. Genie remembers that along the banks of the *bayous* "we could see many kinds of water birds perched on the mangrove trees, the nest of a bald eagle, and an island dense with nesting herons, egrets, cormorants, and pelicans." At low tide they could gather oysters and eat them raw while sitting in the boat; or they'd anchor the boat and trudge across the mud, digging into mud volcanoes to find worms with fancy tentacles on their heads and angel-wing shells with the animals still inside.

At the end of her busy day, Genie returned home to Hera and baby Aya. Hera would be waiting, eager to go swimming. As Genie remembers, "I often brought her home some presents in a pail of sea water: a nudibranch mollusk, with royal-blue and yellow-striped mantle...; an orange starfish with brown warts; and a small purple sea fan I had picked from the reefs." Genie frequently cooked freshly-caught fish for dinner. As she cut open the fish to clean them, she gave an anatomy lesson to Hera and anyone else who happened to be in the kitchen. In the evenings, Ilias studied at his desk while Genie sat comfortably on the sofa, making notes and sketching on a large pad of paper. Occasionally the clatter of their metal garbage can announced the visits of a raccoon Genie named Uncle Yoshi. Otherwise, the nighttime quiet was broken only by the sound of the surf.

In 1956, Genie gave birth to her third child, a son named Themistokles and nicknamed Tak; two years later, she had her last child, Nikolas, known as Niki. She continued diving until just before Niki was born. Genie believes that infants are natural swimmers, and before her children were a week old, she put each of them in deep water in the tub or in the sea. She held them by their chins as they dog-paddled. All of her children learned to swim before they could walk and as they got older, they all went on fishing and diving expeditions with her.

During the first year the Lab was open, Genie's mother, Yumiko, and stepfather, Nobusan, moved their restaurant, Chidori, from New York to Florida. It was the first Japanese restaurant in the state. Located halfway between Genie's house and the Lab, it was a short distance from both. Yumiko and Nobusan were happy to be near their grandchildren and Yumiko was always available to help with the babysitting. Genie could continue her shark research, attend meetings and go on lecture tours knowing her children had

Genie and Ilias Konstantinou with their children Aya, Tak, Hera, and Niki, Sarasota 1959.

the best possible care.

The sea wasn't the only place to go diving. Genie began to explore some of the fresh-water springs near her home. She wanted to collect certain species of fish that swim into the springs from the nearby sea. Florida's springs are famous for their clear water and beautiful underwater scenery. But Warm Mineral Springs and Little Salt Springs, where Genie spent a lot of time, are neither clear nor beautiful. They have very muddy water that is full of sulphur. Black slime covers their walls, reaching down to their sloping bottoms. But Genie wasn't looking at the scenery.

In 1959, William Royal, a diving friend, discovered a prehistoric human bone in Little Salt Springs. He asked Genie to help him explore further. He hoped to find *fossils* and other evidence to prove that Native Americans had lived in Florida thousands of years before. Using scuba equipment, Bill, Genie and other scientists made many dives, deeper than 200 feet. On one dive, Genie went down to 210 feet—without meaning to, she'd set a women's world record for the deepest dive on compressed air. At times it was dangerous work, but the scientists found many traces of Native American life, including several human bones.

In Warm Mineral Springs, the divers explored caves that had once been above ground, but were now covered by water. They found many objects made from deer bones and antlers; also human finger bones and a partly burned log later shown to be 10,000 years old. This was further proof that Native Americans had once lived in those caves. Until then, *anthropologists* had no idea that humans had lived so long ago in Florida.

Bill made a fantastic discovery. He found a complete human skeleton and a human skull next to it. Sticking his finger inside, he found some soft white stuff. Tests showed that it was human brain tissue. The tissue had been preserved 100 feet down in the warm mineral springs water. There was almost no oxygen to make it decompose. Bill and Genie thought it had to be almost 10,000 years old. At first, other scientists doubted that a human brain could have been preserved for so long. Until then, the oldest brain known was less than 2,000 years old. Several years later, radiocarbon dating, a new method of measuring the age of ancient plant and animal remains, proved that the brain from Warm Mineral Springs was about 7,500 years old. News of Bill's discovery spread around the world.

DIVING TECHNIQUES

Breath-hold diving, also called skin and snorkel diving, is the oldest and simplest form of diving. Divers simply hold their breath; most also use a face mask, snorkel tube, and flippers. They usually can go only 30 to 40 feet deep and stay underwater for less than a minute. After resurfacing, a diver using a snorkel tube must be careful to blow the water out of it before taking a breath. Skilled divers, like Siakong, can dive over twice as deep and stay underwater twice as long.

The word "scuba" is an acronym for "self-contained underwater breathing apparatus." Scuba divers wear metal tanks on their backs that hold compressed air which they breathe through a hose. They also wear masks, fins and weighted belts. Recreational scuba divers can safely go down to 130 feet, but a skilled professional like Genie can dive to more than 200 feet. The length of time a scuba diver can spend underwater depends upon the depth of the dive, the diver's physical condition and the amount of exertion while underwater. Tak, Genie's oldest son, proudly tells about a recent dive he took with her in the South Pacific. He was over 200 feet down and she was at least 20 feet below him. "I use more oxygen than she does and at that depth it goes really fast. She's so relaxed down there," he says.

The deeper a diver goes, the greater the water pressure. The increased pressure underwater causes more nitrogen to dissolve in a person's blood than at the surface. (Air, including the compressed air in scuba tanks, contains a large amount of nitrogen). Scuba divers must be careful when rising to the surface. If they come up too rapidly, the extra nitrogen will escape from their blood in bubbles. The nitrogen bubbles can block small blood vessels, causing an extremely painful, and sometimes fatal, condition called "the bends."

6
THE "SHARK LADY"

♦ ♦ ♦ ♦ ♦ ♦ ♦ ♦ ♦ ♦ ♦ ♦ ♦ ♦ ♦ ♦

An interviewer asked Genie what she felt her greatest accomplishment was. She replied, "I think understanding sharks and finding out that they are not stupid, unpredictable, dangerous creatures." Genie has observed them in the wild and in captivity and has dissected them in the lab. In the many years she has devoted to studying sharks, she has learned that they are fascinating, beautiful—and misunderstood. She wants humans to know them better.

Just as people like to shiver in terror at stories about Dracula and the Frankenstein monster, they enjoy scary tales about sea monsters attacking ships and devouring unlucky swimmers. They'd prefer to imagine sharks as savage beasts, rather than creatures who behave according to their natural instincts. Most are *predators* that hunt and eat other animals, but the largest sharks, such as whale and basking sharks, eat only *plankton*. Genie has cut open the stomachs of over 1,600 sharks to study what they eat. She's found the remains of over forty kinds of fish including eels, octopus, sting rays, crabs and other sharks—but never any humans.

Genie has been "bitten" twice by sharks. The first time, she was "bitten" by the teeth of a 12-foot tiger shark. She was on her way to give a lecture at a local high school. Several books and other materials were piled beside her on the car seat. The skeleton jaws of a tiger shark she'd caught were perched on top. When Genie came to a sudden stop for a red light, she stretched out her arm to keep the books from toppling over. The jolt made the shark jaws tumble down and close on her arm, leaving a few minor cuts. The second time Genie was bitten, she had her arm deep inside a female shark in order to deliver an unborn baby. The baby emerged, very much alive, its teeth tightly clamped onto her finger.

The great white shark has an especially bad reputation. It is the largest predatory animal in the world, known as nature's great killing machine. "Man-eater" and "white death" are common nicknames. Many horrible stories have been told about great whites. When people think of them, they imagine a huge mouth opened wide, filled with gleaming rows of razor-sharp teeth. In the famous science fiction classic, *Twenty Thousand Leagues Under the Sea,* Jules Verne describes Captain Nemo's narrow escape from several great whites during an underwater walk. With "silver bellies and huge mouths bristling with teeth" they rush at him out of the darkness. In the movie *Jaws,* a great white hunts, attacks and kills swimmers in the water near the fashionable beach resort town of Amity. It terrorizes vacationers until it is blown up in a duel to the death.

Like other sharks, great whites have special sensors in their heads that detect movement in the water and make it easy to find hidden *prey*. As they close in on a victim, their sensitive noses smell seeping blood, their keen ears, *lateral-line system* and highly specialized electrical sensory system detect sounds and vibrations of splashing and their black eyes pierce the underwater gloom.

Worldwide, great whites attack four or five people a year, perhaps killing one. But no one knows if people are their normal prey. Scientists are currently studying whether great whites even *like* to eat humans. It's possible they don't; when they bite humans, they quickly let go. And if they take a bite, they usually spit it out. Scientists know that great whites prefer to eat animals like seals and whales, which have thick layers of fat. Perhaps humans don't have enough fat to give them the nourishment and energy they get from other animals. In general, scientists say, great whites have been badly misunderstood.

Scientists have some clues but don't always know why sharks occasionally attack people and other times leave them alone. Sharks sometimes attack humans when threatened or provoked. About fifty shark attacks are reported annually worldwide; most of them are not fatal. But it is important for swimmers to be careful in areas known to have sharks. People should never swim if they are bleeding because blood attracts sharks; they should never grab or try to hurt any shark, no matter how harmless it may look.

Genie says that shark attacks are rare and most stories lack good evidence. She has investigated many reported shark attacks and thinks that people make up or exaggerate stories because of the sharks' reputation. "Sharks are as predictable as any animal," she says, "if one takes the time to study and get to know them...it is far safer to swim with these animals than to drive on an average city street or highway."

Forty years ago, when Genie began studying sharks, very little was known about them. Until then, most researchers examined only dead sharks. Their behavior was largely unobserved and barely understood. No one knew why they attacked, how and where they mated, whether they lived alone or in groups, how long they lived or whether they could be trained. Many questions still haven't been answered, but Genie's pioneering studies have provided valuable information.

Sharks appeared on Earth 400 million years ago, 200 million years before the dinosaurs. Modern sharks are very similar to their ancestors and are called "living fossils." In comparison, humans are a young species. The earliest human ancestors appeared about 3 million years ago.

There are over 370 species of sharks that vary greatly in size, behavior and *habitat*. The smallest are 6 inches long and weigh 1 ounce; the largest, whale sharks, are at least 40 feet long and weigh over 20 tons, twice as much as 2 average-sized African elephants.

Sharks are cartilaginous fish. Cartilaginous fish have skeletons made entirely of cartilage which is more flexible than bone; like a human's outer ear, it can be bent and pulled. Sharks, rays, skates and *chimeras* are cartilaginous fish—a group of about 600 species. All other fish are bony fish, an enormous group of about 25,000 species. They have skeletons made of bone, like humans.

The two groups of fish differ in other important ways. Most bony fish have a swim bladder, a gas-filled sac that makes it easy for the fish to stay at a certain depth without sinking. Cartilaginous fish don't have swim bladders. Instead, they have a large liver that is filled with oil. Even

Genie observes a reef shark. Courtesy of Nikolas Konstantinou.

though the oil helps sharks stay afloat, most must swim constantly, even when asleep, so that they don't sink. Sharks also rely on constant swimming to take in oxygen. Like other fish, they must remove oxygen from water in order to stay alive. As sharks swim, seawater flows into their mouths, then passes over their gills. The gills remove oxygen from the water and send it into the sharks' blood.

Another difference is the number of teeth each group has. Unlike bony fish, the jaws of sharks and rays are lined with rows of teeth that are constantly replaced as they wear out. Only one or two rows in front are upright. The teeth in the rows behind lie with their tips pointing toward the back of the shark's mouth. As if they were on a conveyor belt, new teeth move to the front from the space behind the old teeth. A great white shark, for example, can have thousands of teeth during its lifetime. A human has a total of only fifty-two. Sharks even have tiny "teeth" on their skin. These teeth stick out and pro-

SHARK CLASSIFICATION

Class: Chondrichthyes

Subclass: Elasmobranchii

Superorders:

- Squalomorphii; contains three orders, six families and 24 percent of living shark species.
- Galemorphii; contains four orders, twenty-one families and 73 percent of living shark species.
- Squatinormorphii; contains one order, one family, and 3 percent of living shark species.

vide protection, but make a shark's skin feel like sandpaper.

Sharks were called "swimming noses" because scientists used to think that they depended on their sense of smell to hunt prey. It is now known that they depend on their other senses as well. They have excellent hearing, although it is limited to low-pitched sounds, that helps them locate food. And their highly sensitive eyes see well in the oceans' sunless depths. Sharks have a well-earned reputation as the perfect predator.

The anatomy of sharks hasn't changed much since they first appeared on Earth. Based on evidence from fossils and living sharks, their brains have also changed very little. People assumed that sharks were big, stupid eating machines, but Genie suspected this was wrong. She realized it was necessary to work with live sharks in order to observe their behavior. Dissecting dead sharks only provided information about their anatomy.

In the summer of 1958, Lester Aronson, an expert in animal psychology, visited the Lab. "Has anyone ever made a

HUMANS VS. SHARKS

According to records kept since 1990, about fifty humans are attacked by sharks each year; about six humans die of their wounds. But for every human killed by a shark, 2 million sharks are killed by humans. In 1991, over 12 million sharks were killed for food, and it is believed that shark fishing has increased since then.

Commercial shark fishing has become big business. Over the past fifteen years, there has been increasing demand for shark products—shark meat and fins for food, liver oil for vitamins and drugs, cartilage for health-food supplements, hides for leather, teeth for jewelry. Killing sharks for sport has also become popular.

Sharks are vulnerable to overfishing: It takes many years before they are mature enough to reproduce and they have relatively small litters. For example, female great whites take at least ten years to mature and have litters of only seven to nine pups.

Shark conservation is gaining support. Several shark species are becoming endangered and some, such as the great white, are now protected in countries around the world. Growing pollution of the oceans, changes in the food chain and global warming all affect the shark population. Sharks may look and act ferocious, but in the fight for survival against humans, sharks are the big losers.

study of the learning behavior of sharks?" Genie asked him. No one had. She decided to train two large lemon sharks, a male and female, that lived in the shark pen. Dr. Aronson helped design an experiment to prove they could be trained to perform a simple task in order to get food. Laboratory rats and pigeons could be taught to do something similar in only a few days, but Dr. Aronson and Genie thought the sharks might take months.

They made a square wooden target and painted it white. At feeding time, Genie lowered it into the water, a piece of fish dangling from it on a string. When the sharks grabbed the fish, they bumped the target with their noses. The target was connected to a doorbell that rang when the sharks hit the target. Genie hoped they could be trained to connect the sound of the bell with food. She wanted them to learn to bump the target and ring the bell even when food wasn't there. After six weeks of training, she gave them the "big test." Genie lowered the target into the water without any fish attached. At first the sharks swam past it. But on the tenth pass, the male hit it with his nose and the bell rang. Genie immediately dropped some fish into the water. Within three days, both sharks learned they had to bump the target to get their reward.

Then she made the task harder by dropping the food further and further away from the target. The sharks had to learn to bump the target, then swim to the other end of the pen to get the fish. Finally, Genie decided to find out if they would bump the target even if the bell didn't ring. As usual, the male shark charged the target as soon as it was lowered. No bell sounded but Genie dropped the food. The shark learned immediately. The next time he bumped the target without the sound of the bell and got his reward.

Genie continued her learning experiments, working with tiger, bull and nurse sharks, as well as lemon sharks. She proved sharks not only learn simple tasks, but can remember them over several weeks. She designed experiments to see if they could tell the difference between targets of different colors, shapes and patterns. Unfortunately, she lacked the technical equipment needed to test color vision, but it seemed as if the sharks could discriminate between different colors. Her experiments provided new information about sharks and their behavior. The sharks surprised many scientists by showing how easily they could learn simple tasks. They were more intelligent than anyone had thought.

7
A CROWN PRINCE AND MANTA RAYS

♦♦♦♦♦♦♦♦♦♦♦♦♦♦♦

In 1958, Genie and her family moved to a large house in Sarasota, where Ilias had a fast-growing medical practice. The new house had more room and was right on the water, but the move meant Genie's commute to Cape Haze was one hour each way. Over the years, the Lab had grown into a large enterprise. Genie had to spend most of her day dealing with business matters, which left little time for the study and research she loved. In addition to these problems, the Lab would soon have to be moved because it was directly in the path of the planned Intercoastal Waterway.

Then something terrible happened—Yumiko suddenly died. Genie was shocked and grief-stricken to lose the mother who had always loved and supported her. Hera, Aya, Tak and Niki missed their grandmother, who had been a part of their lives for as long as they could remember. Losing Yumiko was especially hard for Nobusan. Since he had never learned to speak English well, he had depended on her to be his interpreter as well as his best friend. After Yumiko's death, he needed Genie's love and attention more than ever. Nobusan was the only father she had ever known, but Genie wasn't sure she could help him adjust to life without Yumiko.

The stress in her family life and the pressure at work overwhelmed Genie. She decided to resign from her job in order to care for her children and stepfather. But the Vanderbilts, along with Ilias and Dr. Breder, suggested that she take a temporary leave of absence and then return part-time. Genie didn't believe that she could run the Lab properly as a part-time director, but said she would try. A year later the Vanderbilts helped her move it to Sarasota, cutting Genie's commute to five minutes. And most important of all, Genie's new housekeeper, Geri, took on Yumiko's babysitting responsibilities. She loved the children, they loved her, and she quickly became part of the family. With Geri to depend on, Genie was soon back at work full-time.

Money was always in short supply at the Lab despite generous funding from the Vanderbilts and grants from the National Science Foundation and other organizations. Genie had gradually cut her own salary back as Ilias began to earn a good living as an orthopedic surgeon. Dr. Heller suggested that she hire an administrator so that she could return to her scientific work. But could the Lab afford to pay someone? Dr. Heller was sure that there would be someone in Sarasota who was bored, capable and willing to volunteer. To Genie's surprise, many talented candidates offered to help. She chose a retired military officer, who took over as business manager and freed her to do research.

Once her life returned to normal, Genie began to travel further away. In 1960, she taught a marine biology course to passengers on an "adventure cruise" in the Bahamas. Her children came along and had their first chance to dive among coral reefs. They were fascinated by the brilliantly colored fish. One day Genie entertained everyone by riding on a giant sea turtle. Clinging to his shell, she rode him back and forth, up and down, steering him through narrow passages in the reefs. The trip to the Bahamas was such fun that Genie included the children on other diving expeditions to the Caribbean islands and Mexico.

In 1964, the family took a special trip to the Middle East. First they stopped in Jerusalem to visit Genie's friends and see the sights. Then they visited the Dead Sea and floated in its warm waters, the saltiest on earth. Finally they reached the Red Sea. Genie took the children diving every day and together they explored the unique coral reefs and a nearby garden eel colony.

There were more than a thousand eels in the colony. They lived in sand burrows on a grassy slope beneath the water. The eels kept the ends of their tails inside their burrows while the rest of their smooth, silver-gray bodies swayed gently above the sand. As they filtered certain plankton from the water to eat, they watched for intruders with their big brown eyes.

A marine biologist friend had asked Genie to bring back a garden eel for the fish collection at Hebrew University in Jerusalem. The eels were almost impossible to catch because they disappeared into the sand whenever a diver approached. Genie tried squirting an irritating chemical into one burrow, hoping to drive the eel out of its hiding place. After several minutes of waiting, she gave up and swam toward the surface. Hera, who was snorkeling, shouted at her mother to look back. A large eel was rising out of its burrow, swaying its head drunkenly from side to side. Genie grabbed the eel,

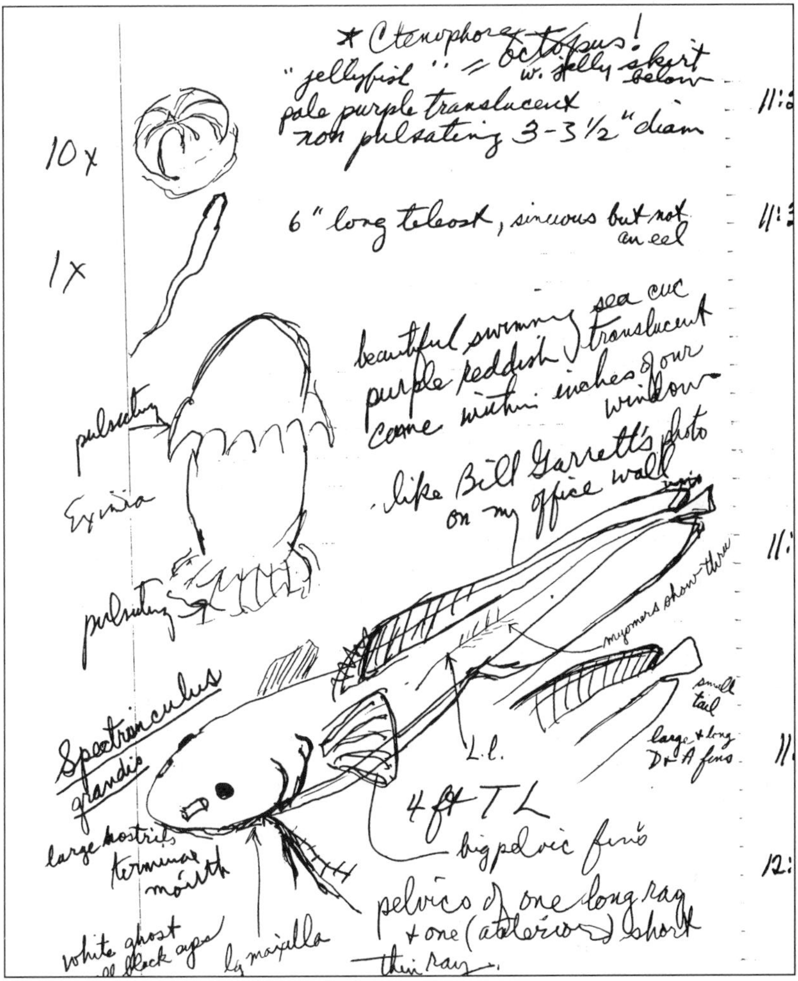

Typical notes and drawings from one of Genie's logbooks show how carefully and minutely she observes and describes what she sees underwater.

pulled it completely free of the sand and brought it to shore.

The eel wasn't the only prize Genie captured during this visit to the Red Sea. She also discovered a new species of sandfish, which she named *Trichonotus nikii* after her youngest son. A "Tricky Niki," Genie's nickname, is slender— about five inches long—and has round shiny eyes. The name *Trichonotus*

WHAT'S IN A NAME?

The scientific name for an organism is part of a classification system invented by the Swedish naturalist Carl Linnaeus in 1753 and still used by biologists today. When a new organism is discovered, the person who has found it chooses a two-part name. The first part identifies the *genus* of the organism and the second part, the species. The genus name is always a noun and the species name is always an adjective. The generic name is always capitalized, but the species name is not. Both names are italicized. Generic names are often descriptive words in Latin or Greek while species names can be more creative. They often honor a person or place, as Genie did when she named a new Red Sea fish *Trichonotus nikii* for her youngest son.

refers to the three feathery, striped plumes on the heads of the males. They spread their plumes like fans to attract females or defend their territory from enemies. But male Tricky Nikis have no problem living right next to garden eels and sharing the same food.

Growing up with a mother like Genie made life an adventure. Her children were used to living with a large assortment of unusual pets and traveling to exotic places on diving trips led by their mother. What they came to accept as normal, others had trouble believing. Once in high school, Niki was required to write a paper about a personal experience with another culture. He described going on a safari across the desert to reach a remote coral reef on the Sinai coast of Egypt. A string of camels carried scuba gear and special cameras along with their riders. These smelly "ships of the desert" seemed especially mean-tempered about having to bear the extra weight of the underwater equipment. Niki's teacher assumed that no one could actually have had such an odd adventure. Deciding he'd dreamed up the whole experience, she gave him an "F" on his paper.

In 1965, Genie visited Japan for the first time. She received a special invitation to meet Crown Prince Akihito. Since the prince was an experienced ichthylogist, Genie decided to bring him a special gift—a young nurse shark. It was trained to ring the bell on a lighted target to get food. The shark rarely made a mistake once it learned to tell the difference between a lighted and an unlighted target. Genie also brought the prince a portable target so that he could watch the shark perform. She carried the little nurse shark with her on the plane to Japan in a container the size of a large hatbox. The stewardesses paid more attention to it than they did to the human passengers.

In Japan, the shark was treated like a royal visitor. A truck carrying a large aquarium met Genie at the Tokyo airport. At the palace, the prince had a special display aquarium set up in a darkened room. He and his guests gathered to watch the performance. A palace servant stood near the aquarium holding a platter of raw lobster slices arranged in the shape of a flower. Each time the shark pushed the correct target, the servant fed it a slice of lobster with a pair of long, elegant chopsticks. After watching the shark perform, Genie had tea and chatted with Akihito, who spoke perfect English. He admitted that, unlike most Japanese people, he didn't like to eat fish. He preferred to collect and study them. He was especially interested in gobies because he could catch them easily by wading in shallow water and using a dip net. To Genie's surprise, he had never dived or even worn a face mask.

For the next few weeks, Genie traveled around Japan and was treated with great respect everywhere. In one city, at a formal dinner given in her honor, she was assigned a "geisha girl" as a companion for the evening. Important male visitors were often honored this way, and the city officials wanted to show the same courtesy to an important woman. The geisha made conversation easy because she spoke English well and was familiar with Genie's research. During dinner, she explained each dish, poured tea and entertained Genie by singing and dancing. The geisha's hospitality and the attention from the Crown Prince made Genie's first trip to Japan very special.

Two years later, Prince Akihito stopped off in Florida to visit Genie on his way home from South America. Although he'd had an exhausting day of travel, he talked about fish with her until midnight, while his wife and attendants struggled to stay awake. Then he surprised Genie by asking for a skin diving lesson the next morning—at 5:30 A.M.! Before sunrise, she met the royal party at their hotel and led them to a nearby beach. The prince looked out of place in his business suit. But at the water's edge, he quickly stripped down to a bathing suit, then waded into the sea to look for gobies. Genie helped him find more than a dozen specimens for his collection. While his security guards waited anxiously on shore, she gave him a quick lesson on how to use a mask and snorkel. Prince Akihito was delighted. He had sneaked in a fish-collecting trip and diving lesson, away from the usual crowds of photographers and reporters.

After the excitement of the royal visit, Genie returned to her work at the Lab. She had begun to study rays, which are related to sharks. Although the manta ray is one of the largest fish found off the western coast of Florida, it is harmless. It eats only plankton and lacks the poisonous tail spines found on most other rays. Mantas are fast, graceful swimmers. They also can leap partly out of the water.

An enormous manta once gave Genie and her assistants an incredible ride. One of the men on the boat hit it with a spear attached to a harpoon gun by a long nylon fishing line. With the spear stuck in its huge body, the manta towed their boat far out to sea. Another boat, which came to help, was tied to Genie's boat. The manta was powerful enough to pull both boats without slowing down. After a few hours, it escaped unharmed. The next day, Genie and her assistants captured another one. But this time, they towed the manta, which weighed over 2,000 pounds. Genie dissected the huge fish on the Sarasota city pier and gave an anatomy lesson to the large crowd of spectators who had gathered to watch.

Genie also studied cow-nosed rays, which gather in large schools and swim close to shore during the summer and early fall. Despite their large size, the schools are difficult to see from shore because the rays stay underwater. Genie first became interested when she was shown an aerial photograph of an enormous school of cow-nosed rays off the coast of Sarasota. She was so fascinated by the size of the school that she counted the fish—there were five to six thousand rays in the photo.

Later, Genie investigated a Coast Guard report that cow-nosed rays were forming huge schools nearby. She thought they might be gathering to mate or give birth or gorge themselves when food was especially plentiful. She and her assistants photographed the rays and caught and dissected many of them. But Genie couldn't find any evidence to support her theories. She spent over a year studying schools of cow-nosed rays, but couldn't solve the mystery. No one has ever been able to explain their strange behavior.

Cape Haze Marine Laboratory continued to expand its research and teaching efforts. Although Genie still enjoyed her work, her personal life was changing. She and Ilias were growing apart. In 1967, they divorced and she returned to New York with the children. Within the next few years, Tak and Niki, then Hera and Aya, moved back to Florida to live with Ilias. They missed Florida's sunny climate and casual lifestyle. "We told Mom, if she'd only move back to Florida, we'd come to live with her," Tak later commented.

During the next few years, Genie acted as an advisor to William Mote, a wealthy businessman who loved the sea and who gradually took control of the Lab. His financial support made it possible to hire Genie's choice for a successor—Perry Gilbert, the shark expert from Cornell. Eventually Cape Haze was renamed the Mote Marine Laboratory, which has continued to grow and to be a highly respected center for marine science. It now leads the world in shark research.

8
SLEEPING SHARKS AND SANDFISH

♦♦♦♦♦♦♦♦♦♦♦♦♦♦♦♦♦♦♦♦

At age forty-five, Genie returned to college teaching. She spent a year as a zoology professor at the City College of New York and lectured at the New England Institute for Medical Research, where John Heller had been director. In 1968, she joined the Department of Zoology at the University of Maryland near Washington, D.C. Despite her busy teaching schedule, she found time to write her second autobiography, *The Lady and the Sharks*, which was published in 1969.

At the University of Maryland, Genie taught courses in ichthyology and zoology. Her enthusiasm for her subject, talent as a speaker and warm, outgoing personality made her one of the most popular teachers on campus. She was generous with her time and knowledge, always ready to encourage students and colleagues. In 1973, the University promoted Genie to a full professorship and she has taught there ever since.

Despite her busy teaching schedule, Genie developed a research program that took her to more than twenty different countries. She invited anyone who was interested in fish and willing to work hard to participate. Her assistants included undergraduate and graduate students, friends, underwater photographers and journalists. Although Hera, Aya, Tak and Niki were living with their father in Florida, they visited their mother during vacations and joined her expeditions whenever they could.

Genie's work attracted the attention of *National Geographic* magazine, which has supported her research and published twelve articles about her major discoveries over the past twenty years. Genie collaborated with staff members and photographers at the magazine and became friends with many of them. Her association with the National Geographic Society also helped make her a television personality. She consulted on and appeared in a 1982 National Geographic special called "The Sharks", which became one of the most popular programs ever shown on public television. It still holds the highest Nielson rating of any TV documentary.

During the 1970s, Genie became famous for her work on the mysterious "sleeping" sharks and the Moses sole, a type of sandfish. Later, much of her research focused on other sandfish like the sand tilefish and the sandperch. She even co-authored a children's book about them called *The Desert Beneath the Sea*.

In 1972, a friend of Genie's, who was a diver and underwater photographer, sent her some pictures of sharks taken in caves off Mexico's Yucatan Peninsula. These large fast-swimming sharks were behaving in a very strange way—they were crowded into the caves and appeared to be sleeping. Usually, fast-swimming sharks need to keep moving in order to "breathe" by pushing oxygen-rich water over their gills. Mexican divers reported seeing several species "sleeping" in

Genie with Sanetaka Kawamata in Hawaii, 1976. Ichthyology is not all work and no play, and Genie has made many friends all over the world.

the caves—lemon, ridge-backed and bull sharks. All of these are normally fast swimmers that can be dangerous to human beings. Yet as long as they stayed in the caves, they remained in a trance-like state and didn't even react when divers handled them.

Genie was curious to see these "sleeping" sharks for herself. With funding from private grants and the Mexican government, she was able to make several trips to the Yucatan

Peninsula. She first saw the sharks in 1973, when she, her student Anita, and Aya swam into a cave where they found sharks lying on the floor in a dazed state. Their eyes were open and they watched the divers, but they didn't react. One female, Genie noticed, "stood still as if for inspection....Her eyes were open....Her mouth opened and closed rhythmically." What was the explanation for this strange behavior?

Genie eventually made ninety-nine dives to visit sharks in three different caves. She and her assistants discovered several interesting features of the caves. Fresh water apparently seeped into them from the mainland. The seepage caused the cave water to be less salty than sea water and changed the *electromagnetic field* inside the caves. Cave water also contained higher amounts of both oxygen and carbon dioxide than did sea water.

Genie noticed that remoras, tiny fish that travel with sharks and eat parasites on their skin and gills, were giving the sharks a careful cleaning in the caves. From her experience raising fish as a child, she remembered that a salt-water bath helped remove parasites from her fresh-water fish. Since sharks live in salt water, the fresh water in the caves probably helped weaken and loosen the parasites from their skin, making them easier for the remoras to remove. Genie thought the sharks might be smart enough to use the caves as "cleaning stations." Some of her assistants wondered if changes in the electromagnetic field in the caves caused the sharks to become "high," inducing a euphoria which felt good, or if the increased amount of carbon dioxide acted as an anesthetic and made them relax. "Sleeping" sharks were later found in caves off the coast of Japan. Genie was so intrigued by the sharks' behavior that she made many trips to both Mexico and Japan to study them. But the mystery of the "sleeping" sharks is still unsolved.

Genie first saw the Moses sole in 1960 when she traveled to the Red Sea to study garden eels. The sole is a flatfish, related to flounders. According to legend, the Moses sole was created when Moses parted the Red Sea. A fish was cut in half. Each half became a new fish—a Moses sole. With its speckled top and white underside, it lies camouflaged on the sandy bottom of the Red Sea, hidden from its enemies. But it has another defense against predators. The first time Genie touched a Moses sole, she noticed a milky, slippery fluid oozing from pores along its fins. The fluid made her fingers feel tingly and numb, and Genie wondered what other effects it might have. A scientific report from 1871 had described the milky fluid, but she didn't have a chance to investigate further until 1972, 100 years later.

When Genie tested the fluid on sea urchins, sea stars and reef fishes, she found that small doses killed these creatures quickly. Then she tried baiting a line with a live Moses sole and placing it in a tank with several sharks. The sharks rushed toward the little fish with their mouths opened wide, but stopped a few feet away. Their jaws seemed frozen open and they jerked quickly away, shaking their heads frantically from side to side. Genie also tried the same experiment in the open sea on free-swimming sharks. She baited an 80-foot shark line with several kinds of fish, including a Moses sole. As she and her assistants snorkeled nearby and watched, hungry sharks devoured all the fish except the sole. Then Genie wiped the skin of the Moses sole with alcohol to remove the poison and threw the fish back into the water. It was instantly swallowed by a shark.

During her years at Cape Haze, Genie was often approached by people who claimed to have invented a shark repellent, but none of them actually worked. She once tested an electrical repellent designed by a group of engineers. When she hung the gadget in the shark pen at the Lab and turned it on, all of the sharks were attracted and came over to investigate the strange box! The poison of the Moses sole, however, turned out to be a powerful shark repellent. Genie found that a tiny amount kept sharks away for more than a day. She originally hoped to turn the poison into an effective commercial shark repellent, but it had a drawback: The chemical compound broke down quickly at room temperature. The poison could not be sold for general use. Later, Genie decided there was no need to develop a shark repellent, since people are much more dangerous to sharks than sharks are to people.

Besides poison, the Moses sole makes another chemical that interests scientists. A student of Genie's discovered that the sole protects itself from its own poison by making an *antidote*. In tests, the antidote also worked against scorpion, bee and snake venom. But it couldn't be used to protect humans because to be effective, it needed to be injected into the blood at the same time as the poison. Later, Genie discovered that the peacock sole, which lives far from the Red Sea near southern Japan, makes a similar poison and antidote.

Genie now spends much of her time studying the sand tilefish that live in both the Red Sea and the Caribbean. The common sand tilefish are ordinary-looking, but behave in some very interesting ways. Most of them live in burrows dug in open, flat, sandy areas near coral reefs. They build mounds on top of their burrows out of broken chunks of coral and almost anything else they can find, including parts of shipwrecks, bits of glass and small pieces of diving equipment.

Some species of sand tilefish live in small groups that include several females and one male protector. If something happens to the male, the largest female grows streamers on her tail, gets larger and changes into a male. The new male takes over as the group's protector.

Another of Genie's current research interests is the sandperch. Like the sand tilefish, it can change from female to male if necessary. Sandperch also live in small groups with one male protecting several females. A sandperch can swivel its eyes independently in any direction, even backwards. If an intruder enters its territory, a male sandperch props itself up on its two front fins and watches, with one eye looking forward and the other sideways.

Female and baby sandperch have small black spots on their faces and are hard to tell apart. But adult males have patterns of stripes on their faces that are so distinctive they can be used to identify individual fish. Genie and her dive team got to know a group of male sandperch from the Red Sea so well that they gave them names—Boom Boom, Fast Freddie, Mr. T, Charlie and Uncle Albert.

9
OCEANS IN DANGER

♦♦♦♦♦♦♦♦♦♦♦♦♦♦♦♦♦

Genie has said, "If I could dive in only one place in the world, I would choose Ras Muhammad." Ras Muhammad is an area of land and water located at the southernmost tip of the Sinai Peninsula. She had heard tantalizing descriptions of it in 1951, but had to wait over ten years to dive there. Over the last twenty-five years, she has led over forty diving expeditions to Ras Muhammad and other Red Sea locations. By the end of the 1970s, she was becoming worried about the increasing threat to its unique marine environment. The Suez Canal had reopened after many years of being shut down by fighting between Egypt and Israel. The reopening of the canal caused an enormous increase in shipping traffic in the Red Sea. With the ships came pollution.

Just west of Ras Muhammad is a shallow channel filled with the underwater roots of mangrove trees and coral formations. The Mangrove Channel was once an ideal place for fish to hatch and grow, but had become choked with plastic, tin cans, oil and tar balls. The steep coral reefs near the edges of the Red Sea had also been damaged by pollution and the beautiful beaches were covered with garbage. When she saw what had happened to her favorite diving spot, Genie felt like crying.

Other changes increased the threat to the reefs. An airport and new roads were built to encourage more tourism, although there were already thousands of visitors each year. Fishermen changed from cotton to stronger nylon fishing lines, that snagged on the fragile reefs and tore up the living coral. Some fishermen dynamited the reefs to kill fish and others dropped heavy anchors which smashed up the coral. A reef that had taken up to 2,000 years to build could be destroyed in only seconds.

Genie was concerned about the future of Ras Muhammad and wanted to do something to help, so she organized the first Ras Muhammad cleanup day. She and a crew of divers filled a truck with garbage collected from the reefs and beaches, but they could scarcely see a difference when they finished. Obviously a one-day effort wasn't enough.

One of Genie's friends was a young diver named Gamal Sadat. In 1980, he introduced her to his father, Anwar Sadat—the president of Egypt. When the president asked about Ras Muhammad, Genie replied that Egypt owned some of the most beautiful, and most endangered, coral reefs in the world. President Sadat promised to create an underwater national park at Ras Muhammad as soon as political discussions with Israel were completed. He even invited Genie to the dedication ceremony that would be held. Over the next few years, the Egyptians and Israelis met to make plans for protecting the Red Sea. But, in a terrorist attack that horrified the world, President Sadat was assassinated. He had not been able to declare Ras Muhammad a national park.

Genie didn't give up. She was making a documentary film about Ras Muhammad when she learned that a fishing tournament was supposed to take place on the reefs. Although she wasn't invited to the opening ceremonies, she went anyway and introduced herself to the head of the tournament. He was Sayed Marei, Sadat's former prime minister. Genie convinced him not only to move the tournament offshore but also to join the Red Sea conservation movement. In 1983, Marei succeeded in having Ras Muhammad declared the first national park in Egypt.

At first, the park wasn't properly staffed and Genie saw little improvement. But when she visited Ras Muhammad in 1989, she found a trained, on-site manager and a team of scuba-diving rangers, who patrolled the park's 75 square miles of land and water. Visitors were forbidden to anchor their boats on the reef and were asked not to feed or interfere with the fish. A new visitors' center with a library and exhibition space had been built. Best of all, the beaches and the Mangrove Channel were once again clean and free of garbage.

When she returned to the United States, Genie attended a meeting of scientists and conservationists at which seven underwater "wonders of the world" were chosen out of twenty-six nominations. Learning about these special places would help people everywhere understand the need to protect marine environments. Ras Muhammad was one of the choices, but Genie worries that even its special status won't protect it from ever-growing numbers of tourists and divers.

Genie's desire to protect marine life wasn't limited to

Hitching a ride on a whale shark. Courtesy of Stan Waterman.

the unusual inhabitants of the Red Sea. In 1981 off the coast of Baja California, she saw her first whale shark and took her first whale-shark ride. The gentle giants fascinated her. Ten years later, she spent a month observing whale sharks at Ningaloo Reef Marine Park off the coast of western Australia. David Doubilet, a famous underwater photographer from *National Geographic* magazine and a good friend, went with her to photograph the rarely seen sharks in their natural environment.

At Ningaloo, cool ocean currents mix with the warm reef waters and create the right conditions to support large numbers of plankton. Genie and David arrived in time for an event that happens once a year—a mass coral-spawning. All of the coral animals in the reef released packets of eggs and sperm at the same time. Genie described it as being "like watching a Mardi Gras in miniature where all the inhabitants were releasing pink and white helium-filled balloons." There were also thousands of tiny red and green worms, which may have been spawning, too. The water around the reef became a nutritious soup, which attracted and nourished huge schools of tiny fish. A few days later, large fish—including dozens of whale sharks of all ages and sizes—arrived to take advantage of the abundant food.

Whale sharks are the largest fish in the world. Adults are typically 30 to 40 feet long. They have thick, rough skin that is dark blue with a pattern of large, white dots and lines. Their coloring is darker above than below, which helps them blend in with their surroundings.

Despite their huge size, there is no evidence that whale sharks eat anything much larger than an anchovy. A whale shark has more than 6,000 teeth in its mouth, but they are tiny and covered by a flap of skin. The shark feeds by swimming through a mass of plankton and slowly moving its head from side to side, "like someone vacuuming in the corners" according to David. When the shark closes its mouth, it forces water out through its gill slits and then swallows the food.

Large, adult whale sharks seem to enjoy being touched, except on their tails, and appear to be unaware of riders. Young whale sharks, which are smaller than adults, seem bothered by human contact. David photographed the sharks and Genie wrote about them for *National Geographic*. Partly because of Genie's magazine articles about whale sharks, thousands of tourists began visiting Ningaloo Reef Marine Park to see them. All inhabitants of the reef are now protected. Visitors can swim with the whale sharks, but are asked not to touch or ride them.

STUDYING FISH TODAY

The techniques Genie uses to study sand tilefish are typical of today's research methods. Most of her equipment can be found in a hardware store. She chooses a section of ocean floor and marks each tilefish mound with a numbered plastic tag. She and her team of divers then measure the mounds with a compass and cotton string that has knots evenly spaced along it. Genie records measurements, observations and drawings on a slate with a waterproof marker or pencil. She records her observations of tile-fish behavior every ten minutes. In order to remain 20 to 60 feet underwater for up to two hours, she wears scuba equipment and usually a thin skinsuit in tropical waters. If the water is cold, she also wears a wetsuit. When Genie returns to the dive boat, she and her assistants put all of their measurements, observations and drawings together to create a scientific record of their findings.

To study deep-ocean fish, Genie uses tiny submarines called *submersibles*. In them she can explore the deepest part of the ocean, up to 12,000 feet below the surface.

Along with other ichthyologists, Genie has changed her research methods as part of the worldwide effort to protect marine life. She rarely uses rotenone and instead of spearing fish, closely observes them in their natural environment. And she, like all other ichthyologists, must obey environmental regulations designed to protect wildlife, such as obtaining permits to collect specimens. She has also given up riding whale sharks and sea turtles and encourages other divers to follow her example. And she frequently represents the United States at international conferences on marine conservation.

10
DEEP-SEA DIVING

♦♦♦♦♦♦♦♦♦♦♦♦♦♦♦

Diving is Genie's passion. She has said, "I love it so much! I think I am a diver first, a scientist second." Her diving career spans developments in diving technology from the simplest to the most advanced. She began by simply holding her breath, then learned helmet diving and went on to become one of the world's best scuba divers. About fifteen years ago, Genie began diving in *submersibles*.

The development of submersibles, vehicles designed for underwater work or exploration, began early in this century. They have many advantages—divers can stay dry and warm and remain at the same pressure as on land. People who go down in diving vehicles don't have to be able to swim and can stay submerged for a much longer time. They don't have to decompress before surfacing. The first manned descent into the ocean's depths was made by Genie's childhood hero, William Beebe. In 1934, he and Otis Barton were dropped into the water off the Bermuda coast, in a vehicle they'd designed called the bathysphere. It was a heavy hollow ball made of iron that was lowered into the ocean by a strong cable. They hung at the end of the cable, 3,028 feet down, unable to land on the bottom. The bathysphere couldn't be controlled. Had the cable broken, it would have sunk to the bottom and stayed there.

Modern submersibles have motors and propellers. They can be manned or unmanned. Unmanned submersibles are known as ROVs, "remotely operated vehicles." ROVs are small robots with mechanical arms and video cameras. They are launched from a ship and controlled through a cable by a technician. Live video images are transmitted through the cable to scientists sitting in front of video screens aboard the ship—almost like being underwater. ROVs are often used to explore the depths between 180 feet, where most scuba divers can't safely go, and 800 feet, where manned submersibles are used. They are much cheaper to use than manned submersibles.

Manned submersibles are operated by a pilot and have room for at least two passengers, usually scientists. Their strong hulls are built to withstand water pressure at extreme depths, some to more than 20,000 feet—over 3.75 miles! Manned submersibles carry their own air supply; they have cameras and floodlights so passengers can take photographs in the dark waters near the ocean floor. Some have mechanical arms attached outside that can pick up objects.

Inside, racks of instrument panels surround the pilot. Passengers are crammed into a compartment, usually ball-shaped, that measures 6 feet across. They share their space with lunch boxes, cameras, computers, blankets, fire extinguishers and flashlights. Each passenger sees the sights of the submarine world through a viewport, a window made of plastic. The silence is broken by the beeps, pings, hums, gurgles

Genie with her friend and colleague Emory Kristof as they prepare for a deep-sea dive in the Russian submersible Mir 1. Photo by Dr. Andreas Rechnitzer

and thunks of instruments, valves and pumps, and the hiss of oxygen flowing into the compartment. The complicated technology makes manned submersibles uncomfortable to ride in and expensive to operate, but they have revolutionized undersea exploration.

Genie entered a new phase in her career when she began diving in submersibles. From 1987 to 1990, she was the chief scientist on seventy-one submersible dives with the Beebe Project. The idea for the project came from Emory Kristof, a wildlife photographer with *National Geographic* who had been a friend and diving companion of Genie's for several years. In addition to wildlife, Emory also photographed famous shipwrecks. For his excellent work on a photo-essay about the *Titanic* and other famous shipwrecks, the National

Geographic Society continued to support his deep-sea work, including the Beebe Project. When Genie agreed to become chief scientist, she contributed not only her expertise about fish but also grant money to help fund the expensive dives.

The Beebe Project dives ranged from 1,000 to 12,000 feet deep and took place all over the world—off the coasts of California, Bermuda (where Beebe began deep-sea exploration), the Bahamas, the Cayman Islands and in Suruga Bay, Japan. The submersibles, too, came from all over: U.S. *Alvin*, owned by the United States Navy; *Pisces VI* from Canada; *Nautile* from France; and the Russian submersible *Mir 1*.

Before using submersibles, Emory had to take photographs of deep-sea life by lowering special cameras, with bait attached, into the oceans. The cameras were set to take photographs at intervals, but often the pictures were unclear or the fish had disappeared when the shutters opened. Emory realized that he would get better results by taking the pictures himself, so he invented a technique for deep-sea photography. He devised a way to attach cameras onto a submersible. To attract his quarry, a bait cage filled with chunks of dead fish was carried down by the submersible's mechanical arm and dropped when it reached the ocean floor.

When the submersible first landed, its noise and lights seemed to frighten the sea creatures. But after the pilot turned off the lights and backed the vehicle away, most of them returned to eat the bait. Emory and Genie sat quietly for long periods, observing and videotaping everything within range. Genie also drew sketches and took notes, just as she had during her other observations, even though she was thousands of feet underwater.

They saw deep-sea sharks and other strange creatures

never before seen in their natural habitat: a huge 15-foot elephant-ear sponge, the biggest sponge ever seen; tiny, 6-inch-long lantern sharks; cookie-cutter sharks that bite deep into their victims and pull out a round plug of flesh; deep-sea gulpers with huge heads and hinged mouths; and a hooded octopus with fins on its head that looked like gigantic ears.

In 1990, Genie and a team from *National Geographic* made a series of ten dives in Suruga Bay, Japan. What makes Suruga Bay special are its waters, which drop to 8,000 feet just a few miles offshore, and the deep-sea creatures that live there. On one of the dives, three team members had an unexpected encounter at 4,000 feet. "The largest creature ever seen in the deep sea lumbered in front of the view ports of the submersible *Nautile* on September 13, 1989." A Pacific sleeper shark crashed into the bait cage and buried it in the mud. The cameraman said, "We saw a fish bump into [the] wall, and then the wall moved. The sub shook...All we could think was holy mackerel!" Even though she hadn't been on the dive, Genie later estimated that the sleeper shark was over 23 feet long, almost as long as the 26-foot submersible.

Genie's longest dive was in the Russian *Mir 1*, off the coast of Bermuda. It lasted seventeen-and-a-half hours, but the time seemed to fly by. She counted twenty-one sharks, including several species of the primitive six-gill shark. Her deepest dive, in the U.S. *Alvin*, went down to 12,000 feet. Chimeras, whose slowly flapping fins made them seem like birds in flight, passed in front of the viewports. Since they are relatives of sharks, Genie wonders why there are no sharks at that depth and why sharks are rarely, if ever, seen below 7,000 feet. She also wonders about megamouths—huge, plankton-eating sharks that live in the deep sea. Only eleven have been caught; maybe divers will discover where they live and someday megamouths will be better understood.

Genie thinks that diving in a deep-sea submersible is as exciting as going up into space. She has fulfilled her longtime dream of exploring the deepest waters of the ocean. She has seen the sights William Beebe described when he dove in the bathysphere; she has dived deeper and seen much more. But Genie's quest continues. She wants to find out what other fish and animals live in the oceans' depths. She wants to know more about the fish that live nearer the surface. She will share what she learns with students, other scientists and the public. She will have more adventures. And she will keep on diving.

EPILOGUE

Most people over seventy-five years old slow down—but not Eugenie Clark. Although she officially retired from the University of Maryland in 1992, she still has an office and lab on campus. Genie also has an office and lab at Mote Marine Laboratory in Sarasota, Florida, which she shares with her old friend and colleague Perry Gilbert. Once a year, she teaches a popular honors course at the university called "Sea Monsters and Deep-Sea Sharks." The first class begins with a film in which Genie rides a whale shark. Before the class ends, she explains why no one should ride them anymore.

In an interview, Genie told a story about a recent experience. She was snorkeling with a friend near La Paz, Mexico. She had just announced to a large group of people that the whale sharks should not be disturbed or ridden, when a big shark surfaced directly beneath the two of them. The two women held onto each other so they wouldn't be tempted to grab the shark. But it seemed to be unaware of them. It rose up under them so they were lying on its head, then flipped them backwards as they struggled not to give in and take a ride. Genie was determined to follow her own advice—look but don't touch.

Genie has made friends from all over the world. In 1997, she returned to the South Sea Islands for the first time in almost fifty years. She wanted to see Siakong's sister,

Uredekl, but couldn't make the trip to Kayangel. To her surprise, she learned that Uredekl was ill and had been brought to the main island of Koror for medical treatment. Genie wanted to visit. As she tells it, "It was a wonderful, tearful, touching and miraculous reunion. They only told Uredekl that she had a visitor, but the moment I walked in and she saw me she shouted, 'Eugenie Clark!'" Uredekl died soon after.

Genie gives lectures all over the world and leads several research expeditions each year to some of her favorite places like the Red Sea, Japan, Mexico and the Caribbean. She has discovered and named eleven new species and is currently studying sandfish and large deep-sea sharks. She is also investigating a mysterious eel-like fish that burrows under coral reefs.

During almost fifty years as an ichthyologist, Genie has written more than 150 articles and three books. Most of the articles were published in scientific journals, but some of them, as well as her books, were written for the general public. Through lively writing and vivid descriptions she brings the wonders of the underwater world to her many readers. In addition, she has consulted on twenty-four television specials and appeared in many of them, including a spectacular IMAX film, *In Search of the Great Sharks*.

Genie's contributions to marine science have earned her dozens of honors and awards. They include a gold medal from the Society of Women Geographers, two medals from the Egyptian government, a special award from the National Geographic Society and an international film festival award for the best nature film, *Ras Mohammed National Marine Park*. Genie has also been honored by having four new fish species named after her: *Callogobius clarki*, *Sticharium clarkae*, *Enneapterygius clarkae* and *Atrobucca geniae*.

Genie and Henry on their wedding day

Recently Genie married an old friend, Henry Yoshinobu Kon, and spends most of her time in Sarasota, Florida with him. She's proud of her children's accomplishments and sees them often. Hera is working on a Ph.D.in marine biology and lives in Florida with her husband. Aya is a commercial airline pilot and competes in horse shows. She and her son Eli, Genie's only grandchild, also live in Florida. Eli is only nine, but is already an excellent swimmer and diver. Genie adores him and likes to babysit. She takes him along on diving trips whenever possible. Tak, who also lives in Florida, is an award-winning photographer and Niki specializes in underwater photography and film production.

Genie is modest about her accomplishments. She believes that she has corrected misunderstandings about sharks, added to what is known about fish and inspired young people, espe-

Genie often brought her children on diving trips. Her older son Tak went to the Yucatan with her in 1975 to study sleeping sharks.

cially girls, to study science. When asked if it's difficult to be a woman in a male-dominated field, she replied that it was sometimes hard to get people to take her seriously at first; but whenever she succeeded at what was considered man's work, she felt she got more credit because she was a woman.

Genie loves her life and wouldn't change anything about it. She enjoys thinking about all she's done, but doesn't want to live in the past. The future is what excites this extraordinary woman. As she says, "What's done is done. I don't like repetitions....There is still so much to learn and experience."

SOURCES CITED

p. 6 Eugenie Clark, *Lady with a Spear* (New York: Harper & Brothers, 1953), p. 4.

p. 9 Eugene K. Balon, "The life and work of Eugenie Clark: devoted to diving and science," *Environmental Biology of Fishes* 89 (1994), p. 89.

p. 10 Clark, *Lady with a Spear*, p. 15.

p. 11 Ibid. p. 13.

p. 12 Ann McGovern and Eugenie Clark, *Adventures of the Shark Lady: Eugenie Clark Around the World* (New York: Scholastic, 1999), p. 78.

p. 13 Encyclopedia International, 11th ed., s.v. "Pearl Harbor."

p. 14 Balon, pp. 122-23 .

p. 17 Clark, *Lady with a Spear*, p. 24.

p. 20 *The Economic Report of the President of the United States,* 1973.

p. 29 Clark, *Lady with a Spear*, p. 109.

p. 32 Ibid., p. 162.

p. 33 Ibid., p. 165.

p. 33 Ibid., p. 191.

p. 39 Ibid., p. 195.

p. 41 Ibid., p. 179.

p. 50 Eugenie Clark, *The Lady and the Sharks* (Sarasota: Mote Marine Laboratory, 1995), p. 38.

p. 50 Ibid., p. 29.

p. 54 Tak Konstantinou, online interview by the authors, 17 February 1999.

p. 55 John Stein, "Interview: Eugenie Clark," *Omni* 4 (June 1982), p. 118.

p. 56 William J. Broad, "Shark Is Efficient Killer, but Picky Eater," *New York Times* (July 7, 1997), p.C1.

p. 57 Eugenie Clark, "Sharks: Magnificent and Misunderstood," *National Geographic* (August 1981), p. 178.

p. 60 Clark, *The Lady and the Sharks*, p. 95.

p. 72 Madeleine Lundberg, "Eugenie Clark: Shark Tamer," *Ms*. 8 (August 1979), p. 15.

p. 76 Lisa Yount, *Contemporary Women Scientists* (New York: Facts on File, 1994), p. 66.

p. 80 Eugenie Clark, "Expedition: Red Sea," *Sea Frontiers* (October 1992), p. 21.

p. 83 Eugenie Clark, "Whale Sharks: Gentle Monsters of the Deep," *National Geographic* (December 1992), p. 127.

p. 84 Ibid., p .131.

p. 86 Richard F. Burns, "Dr. Eugenie Clark—The 'Shark Lady'," *Sea Technology* (February 1992), p. 77.

p. 90 David Doubilet, "Suruga Bay. In the Shadow of Mount Fuji," *National Geographic* 178 (October 1990), p.11.

p. 94 Eugenie Clark's addition to manuscript

p. 96 Balon, p. 124.

GLOSSARY

anatomy—the science that deals with the structure of living organisms

anthropologist—a scientist who studies humankind

antidote—a remedy that counteracts the effects of a poison

bathysphere—a strongly built steel diving sphere used for deep-sea observation

bayou—a creek or small river that is marshy or slow-moving

chimera—a deep-water, cartilaginous fish related to sharks and rays

camouflage behavior—an organism's ability to blend in with its surroundings because of its color, pattern, shape or a combination of these characteristics

dissect—to separate a dead organism into pieces for scientific examination

dissertation—a paper written on an original subject that is required for a doctorate

dorsal fin—a fin located on the back of a fish

electromagnetic field—the space around an object that has been altered by the combined forces of electricity and magnetism

endocrinology—the study of specialized glands in the body, such as the thyroid, that produce hormones

fossil—any evidence of an extinct organism that is preserved in the earth's crust

habitat—the place where a plant or animal normally lives and grows

hybrid—an offspring of two animals or plants of different species

ichthyology—the study of fishes

invertebrate—an animal without a backbone

lateral-line system—in sharks, a series of fluid-filled canals that contain tiny hairlike receptors sensitive to vibrations and pressure changes

nautilus—a marine animal related to octopuses and squids that has a spiral, multi-chambered shell with a pearly inside

nudibranch—a sea snail that has no external shell

pectoral fin—either of the fins of a fish that correspond to the front limbs of a four-legged animal

physiology—the study of the functions and activities of living organisms

plankton—usually tiny, free-floating aquatic plants and animals that drift with the current

plectognaths—a group of nine families of bony fish that swim slowly, have small gill openings and live mostly in tropical ocean waters near coral reefs

predator—an animal that kills and eats other animals

prey—an animal eaten by a predator

scuba—the acronym for "self-contained underwater breathing apparatus"

species—a group of related organisms that can mate and produce offspring

submersible—a small manned or unmanned vehicle used for underwater work or exploration, especially in the deep ocean

tide pool—a pool of water remaining on a reef, shore or beach after the tide has receded

vertebrate—an animal with a backbone

zoology—the study of animals

BIBLIOGRAPHY

Balon, Eugene K. "The life and work of Eugenie Clark: devoted to diving and science." *Environmental Biology of Fishes* 41(1994): 89-114.

Broad, William J. "Shark Is Efficient Killer, but Picky Eater." *New York Times* (July 15, 1997): C1-2.

Burns, Richard F. "Dr. Eugenie Clark—The 'Shark Lady'" *Sea Technology* (February 1992): 77.

"Clark, Eugenie." *Current Biography* (1953): 120-22.

Clark, Eugenie. "Expedition: Red Sea." *Sea Frontiers* (October 1992): 20-28.

Clark, Eugenie. Interview by the authors. 9 November 1998.

Clark, Eugenie. *The Lady and the Sharks*. Sarasota, Florida: Mote Marine Laboratory, 1995.

Clark, Eugenie. *Lady with a Spear*. New York: Harper & Brothers, 1953.

Clark, Eugenie. "Sharks: Magnificent and Misunderstood." *National Geographic* (August 1981): 148-86.

Clark, Eugenie. "Whale Sharks: Gentle Monsters of the Deep." *National Geographic* (December 1992): 123-38.

Clark, Eugenie and Emory Kristof. "Sharks at 2,000 Feet." *National Geographic* 170 (November 1986): 680-91.

Doubilet, David. "Suruga Bay. In the Shadow of Mount Fuji." *National Geographic* 178 (October 1990): 2+.

Facklam, Marjorie. *Wild Animals, Gentle Women*. New York: Harcourt Brace Jovanovich, 1978.

Hauser, Hillary. *Women in Sports: Scuba Diving*. New York: Harvey House, 1976.

Konstantinou, Niki. Online interview by the authors. 22 February 1999.

Konstantinou, Tak. Online interview by the authors. 17 February 1999.

La Bastille, Anne. "Scientist in a Wetsuit." *Oceans* (September 1981): 44-50.

Lundberg, Madeleine. "Eugenie Clark: Shark Tamer." *Ms.* 8(August 1979): 12+.

Marx, Robert. *Sea Fever*. New York: Doubleday, 1972.

McGovern, Ann. *Shark Lady: True Adventures of Eugenie Clark.* New York: Four Winds Press, 1978.

McGovern, Ann. *Adventures of the Shark Lady: Eugenie Clark Around the World.* New York: Scholastic, 1999.

McGovern, Ann and Eugenie Clark. *The Desert Beneath the Sea*. New York: Scholastic, 1991

Shanks, Judith W. "A field trip underwater with the renowned 'Shark Lady'." *Sea Frontiers* 39 (November-December 1993): 40+.

Sharks & Rays. [Alexandria, Va.]: Time-Life Books, 1997.

Springer, Victor G. and Joy P. Gold. *Sharks in Question*. Washington, D.C.: Smithsonian Institution Press, 1989.

Stein, John. "Interview: Eugenie Clark." *Omni* 4 (June 1982):94+.

Van Dover, Cindy. *The Octopus's Garden: Hydrothermal Vents and Other Mysteries of the Deep Sea*. Redding, Mass.: Addison-Wesley, 1996.

Yount, Lisa. *Contemporary Women Scientists*. New York: Facts on File, 1994.

ACKNOWLEDGMENTS

The authors wish to give special thanks to Bev Rodgerson, Genie's assistant at the University of Maryland. She was endlessly patient about obtaining answers to our questions and facilitating contact with Genie, her family and friends.

In telephone conversations and e-mail correspondence, Aya, Niki, and Tak Konstantinou provided us with a unique perspective on their mother. In addition, Niki generously made several of his professional photos available. Hera Konstantinou spent many hours locating otherwise unavailable family photos in her collection. Genie's colleagues at National Geographic also gave us invaluable assistance: Emory Kristof talked to us at length about Genie as a colleague and friend and offered access to his photos; and Bruce Hunter went out of his way to search his collection for photos we might be able to use. Genie's good friend, professional photographer Ruth Petzold, generously provided the family photo on the book's back jacket. Thanks also to Lisa Palmer at the National Museum of Natural History and Lara Papi at the National Aquarium for their assistance. And finally, plaudits to our publisher and editor, Diantha Thorpe, for her careful attention to our manuscript.

INDEX